Cytology of Transudates and Exudates

Monographs in Clinical Cytology

Vol. 5

Editor
GEORGE L. WIED, Chicago, Ill.

Co-Editors
EMMERICH VON HAAM, Columbus, Ohio; LEOPOLD G. KOSS, New York,
N.Y.; JAMES W. REAGAN, Cleveland, Ohio

S. Karger · Basel · München · Paris · London · New York · Sydney

Cytology of Transudates and Exudates

EMMERICH VON HAAM

Department of Pathology, Ohio State University, Columbus, Ohio

200 figures and 8 tables, 1977

S. Karger · Basel · München · Paris · London · New York · Sydney

Monographs in Clinical Cytology

Vol. 1: JAMES W. REAGAN and ALAN B.P. NG (Cleveland, Ohio): The Cells of Uterine Adenocarcinoma. VIII + 144 p., 97 fig., 4 tab., 2nd, revised edition, 1973.
ISBN 3-8055-1621-5

Vol. 2: JOHN K. FROST (Baltimore, Md.): The Cell in Health and Disease. An Evaluation of Cellular Morphologic Expression of Biologic Behavior. XII + 148 p., 81 fig., 2 diagrams, 15 charts, 1969.
ISBN 3-8055-0440-3

Vol. 3: STANLEY F. PATTEN, JR. (Rochester, N.Y.): Diagnostic Cytology of the Uterine Cervix. X + 210 p., 190 fig., 1 tab., 1969 (out of print, 2nd, revised edition in preparation).
ISBN 3-8055-0441-1

Vol. 4: J. ZAJICEK (Stockholm): Aspiration Biopsy Cytology. Part 1: Cytology of Supra-diaphragmatic Organs. XII + 212 p., 141 fig., 16 tab., 5 clp., 1974.
ISBN 3-8055-1407-7

Cataloging in Publication
von Haam, Emmerich, 1903
Cytology of transudates and exudates
Emmerich von Haam. Basel, New York, Karger, 1977.
(Monographs in clinical cytology, v. 5)
1. Exudates and Transudates – cytology 2. Neoplasms – pathology I. Title II. Series
W1 MO567KF v. 5/QZ 200 V946c
ISBN 3-8055-2625-3

Contents

I. Introductory Remarks

This book is one of a series of monographs in clinical cytology originally published by Karger in Switzerland. The purpose of this series of books is to present to the cytologist and cytopathologist in concise form the latest advances in the various fields of clinical cytology.

Cytology of transudates and exudates has been studied since 1867 when LÜCKE and KLEBS first described the appearance of malignant cells in pleural exudates. Since then, this observation has been confirmed by many authors who demonstrated the existence of cancer cells by many methods. One of the latest monographs in this field appeared in 1968 by SPRIGGS and BODDINGTON [189], who published a second and enlarged edition of the original monograph of SPRIGGS published in 1957. The new edition contains 195 photographs, partially in color, and a list of 231 references. Because of the completeness of the chapter on the history of the cytodiagnosis of tissue fluids, this author feels it unnecessary to repeat this complete historical review, but prefers to bring the literature up from 1966 to 1975. Chapter II has been devoted to this task and contains 317 references. Chapter III refers briefly to the anatomy, histology, and normal cytology of the four large serous cavities of our body: the pericardial cavity, both pleural cavities, and one peritoneal cavity [74]. We have omitted the cytology of the synovial cavities. Chapter IV discusses the several cellular components of transudates, while chapter V describes the characteristics of exudates of inflammatory conditions [70]. Chapter VI describes the characteristics of effusions produced by malignant tumors with special emphasis on the description of metastatic malignant cells. Chapter VII includes a description of the effusions caused by leukemias and lymphomas. Chapter VIII describes the effect of chemotherapy and radiotherapy upon the benign and malignant cells of effusions and exudates.

This book is illustrated by 200 microphotographs of cells from the various types of transudates and exudates with brief legends. The microphotographs are arranged four to each page, similar to the monograph by REAGAN and NG [160]. They all present pictures of cells stained by the Papanicolaou method or cells from cell blocks stained with hematoxylin and eosin. The magnifications vary from 700 to 1,600. The exact description of the cells can be found in the text adjoining the pictures.

Chapter IX presents an analytical study of our own material which includes 1,054 cases of exudates and transudates examined during the period from 1962 to 1974. It will also include a discussion of the accuracy of our cytological diagnoses [84, 85].

Chapter X briefly describes the methods used in our laboratory for the examination of transudates and exudates with critical comments on the methods used by other authors of this field.

Because of the confined subject matter of this monograph, there will be no subject index.

II. Review of the Recent Literature

The older literature abounds with descriptions of the various cell types found in transudates and exudates of the serous cavities. At first, the problem of recognition of cancer cells seemed most important, but later attention was also paid to inflammatory and congestive types of effusions and a classification of the various types was attempted by WIDAL and RAVAUT [204]. GRUENZE [83] emphasizes that no uniform cell formula exists for the recognition of any specific disease. It appears simply sufficient to lay down certain estimated relationships of the cell distributions for each disease. An additional difficulty is the tremendous morphological variations between the mesothelium of the serous membranes, the histiocytes, the lymphocytes, and the small tumor cells.

SPRIGGS and BODDINGTON [189] wrote an excellent review on the cyto-diagnosis of serous fluid which cites the monographs of QUENSEL [157, 158] as standard works covering the important work of the 19th century. At that time, most of the smears were stained by the Giemsa-Romanowski method, while a few authors preferred supravital dyes, phase contrast methods, and the histological techniques of cell block preparations [205]. With the introductions of the stain by PAPANICOLAOU, a new method was added which gained quick popularity. SPRIGGS [190] finds the results obtained by this technique rather unreliable, particularly because of the level of false-positive reports which to him are completely unacceptable.

Four worthwhile articles appeared in the *Acta Cytologica* in 1966. One, that by JOHNSON [101], was already reviewed by SPRIGGS and BODDINGTON [189]. It carried a large number of false-negative reports (42%). Another article was written by KONIKOV *et al.* [110]. It dealt with 158 patients who had malignant cells in their effusions and were followed until death. The average survival time for carcinomas which appeared in effusions was 3.3 months, while patients with lymphomas lived 2 months. The longest survival time in their series occurred in patients with metastatic carcinoma of the breast and the ovary, who often survived 2 years or longer. CARDOZO [32] wrote a critical evaluation of 3,000 cytologic analyses from pleural fluids, ascitic fluids, and pericardial fluids. 23% of his cancer patients had negative reports and could be classified as false-negatives. A class V positive report showed an accuracy of 99.9%. CARDOZO [32] objects to reports indicating 'suspicious of malignancy', since they are of no value to the clinician. He also feels that suspicious reports should not be counted amongst the positive reports. In his series, the examination of pleural fluids was ten times as frequent as ascitic fluids. In 9% of his pleural fluids, no malignant cells could be encountered, although the patient definitely had tumor metastases. BERGE and HELLSTEN [13] compared the results obtained by pathologists and cytologists in the cytologic diagnosis of cancer cells. No false-positive reports were found by either one. The pathologist demonstrated tumor cells in 64% of the 174 cases, while the cytologist demonstrated them in 75%. Those figures include both positive and suspicious reports.

Interesting case reports on pericardial effusions in rheumatoid disease were written by BERGER and SECKLER [14], LATHAM [114], and HOLLANDER et al. [94]. The diagnosis was primarily made from the positive latex reaction and the occurrence of rheumatoid arthritis cells (RA cells) and confirmed by needle biopsy of the pleura. Other interesting case reports include the occurrence of pericardial effusions in cases of hypopituitarism by STEPHANS et al. [191], the occurrence of pericardial effusions in a case of amoebic pericarditis by SOLAN [186], the occurrence of isolated or primary chyloperi-cardium by KNIGHT [109] and HUDSPETH and MILLER [95] and MILLER et al. [142], and the occurrence of pericardial effusions as a sequel to myocardial infarctions by DEMPSEY et al. [55]. Severe thyroid insufficiency has been de-scribed as a cause of pericardial effusions by DEGEORGES et al. [52] and BURGERMAN [26]. Early ascites with or without pleural effusions have been reported in pancreatitis by GIRARD et al. [79] and SARRAZIN et al. [171].

The importance of amylase determinations in the serous fluids has been stressed by MODAI et al. [145], who published three cases. The determination of other enzyme systems for the diagnosis of pleural or peritoneal effusions has been stressed by BERNARD et al. [15]. Ascites can also appear as an early symptom in idiopathic total portal vein thrombosis as reported by WHITE [202]. A contribution to the problem of eosinophilic pleural effusion was made by CONTINO and VANCE [41], which again stressed the multiple etiolo-gical factors for this condition.

Efforts to improve the cytodiagnosis of effusions were published by MAVROMATIS [128, 129] and MASIN and MASIN [124]. MASIN and MASIN [124] used the desorption technique for a differentiation of mesothelial cells, which desorb easily, from malignant cells, which do not desorb at all. The procedure is simple. The stained cells are shaken in the suspension of 3% hydrogen peroxide and then examined. Benign mesothelial cells and histiocytes will appear pale as compared to malignant cells, which retain their strong color. MAVROMATIS [128, 129] used the appearance of PAS-positive intracyto-plasmic granules or droplets for the positive identification of benign meso-thelial cells, while histiocytes and malignant cells do not have those granules. This test, which was used widely by CEELEN [38] and FOOT [68], really has not proven very satisfactory since many typical mesothelial cells also do not contain PAS-positive granules.

The place of acridine orange fluorescence in the cytology of serous effusions was studied by PALIARD et al. [154], while CLARKSON et al. [40] studied the kinetics of proliferation of cancer cells in neoplastic effusions.

In 1967, KERN [104] took issue to CARDOZO's [32] article in *Acta Cyto-logica* by questioning the efficiency of his staining technique and the high number of mesotheliomas which were found in 10% of all cancer patients with pleural effusions. Current concepts of peritonitis were discussed by DAVIS [47]. He stressed rapid correction of fluid and electrolyte shifts fol-lowed by prompt administration of antibiotics as the most urgent measures before surgery should erradicate the focus or cause of the peritonitis. An interesting case of *Candida* peritonitis was reported by SHAPIRO [178], which was successfully treated by Amphotericin B. A patient with pseudo-Meig's syndrome developed peritoneal and pleural effusions and cytological ex-amination revealed a benign mucinous cystadenitis [183]. Ascites secondary to benign pancreatic disease is quite a frequent occurrence according to the report of CAMERON et al. [30]. The authors consider ascites and pancreatitis a definite clinical entity, which is usually overlooked because of the alcoholic history of most of the patients. The differentiation of this disorder from ascites in hepatic cirrhosis can be made on the basis of the high amylase and the high protein content in the pancreatic ascites. Surgical pancreatic drain-age seems to be curative. SVANE [193] reported a case of acute pancreatitis which developed tremendous ascites with a bilateral pleural exudate. The ascites was characterized by marked eosinophilia. A case of pleural endo-metriosis with hemothorax was reported by YEH [212]. It was associated with endometriosis in the bronchopulmonary tissues and widespread abdominal endometriosis. Four cases of malignant mesothelioma of the pleura were reported by ALP and KARADENIZ [2] of which three proved invasive, while one could be removed surgically. A case of bacterial endocarditis com-plicated by massive hemorrhagic pericardial effusion and cardiac tamponade was reported by ROSE et al. [168]. Pericardial effusions appear also in patients with alcoholic myocardopathy [106] and in one patient with primary chylopericardium [65]. HIRSCH et al. [92] report a case of metastatic breast cancer in a 34-year-old female, which persisted for 6 years and had several remissions following radiation therapy and chemotherapy. A large hydro-pericardium was reported as a fatal complication in a case of Marfan's syndrome by MATSUI and ITO [126]. The diagnostic significance of lympho-cytes in pleural effusions was stressed by YAM [211], who found that 73% of lymphocytic effusions were found in cases of tuberculosis, while the re-maining cases were due to lymphoma, carcinoma, and chronic pulmonary infection. The diagnostic value of paracentesis for the management of the acute abdomen was recommended by MORRIS [149]. Its main usefulness is in intra-abdominal trauma, primary peritonitis, and acute pancreatitis. Study of the peritoneal cytology by puncture of the Douglas space or in per-celioscopic samples may lead to the recognition of ovarian disorders or in-flammatory processes of the female sex organs according to DUPRÉ-FROMENT et al. [58]. He recommends this diagnostic methods as a part of the routine follow-up after surgery for carcinoma of the female genital tract. The deter-mination of karyotypes in cells from malignant effusions has been studied by SANDBERG et al. [170]. Cancerous effusions show a wide range in the number of chromosomes ranging from 35 to more than 100 chromosomes. No two effusions show a similar karyotype. Marker chromosomes are very frequent, but are not related to the total number of chromosomes. The karyotypes persist in spite of therapy and remain the same after recurrence. The therapy of exudates caused by cancer can be effectively handled by the administration of radioactive substances as the studies of SWENSON et al. [194] and ARIEL et al. [4, 5] indicate. Persistent exudates in cancer patients carry an ominous prognosis with an average duration of life after establish-ing the diagnosis lasting about 6 months. Ascites can form pleural effusions and vice versa, either through transfusions through the diaphragmitself, or through the diaphragmatic lymphatics.

In 1968, we find the rebuttal of CARDOZO [33] to the Editorial Letter in *Acta Cytologica* from KERN [104]. He explained the high number of meso-theliomas in his series by the fact that those represent mostly referral cases, and he divided mesotheliomas into well-differentiated, moderately well-differentiated, and highly undifferentiated types. His false-positive results are considered as the percentage of erroneous reports on the sample at the moment that both clinical diagnosis and therapy have yet to be established.

DEBRUX et al. [51] discuss the cytology of peritoneal fluids sampled by coelioscopy or cul-de-sac puncture. On the material of 250 samples, he recognized four types of cytologic patterns during the course of the menstrual cycle. The preovulatory type is generally marked by a hemorrhagic exudate with mesothelial hyperplasia possibly in relation to hyperestrogenism. The postovulatory or early secretory type shows peritoneal irritation with exfoliation and transformation of mesothelial cells. The advanced secretory or premenstrual type shows patches of hyperplastic mesothelial cells and absence of inflammatory elements. The menstrual and paramenstrual type is hemorrhagic and very cellular. Large clumps of mesothelial cells appear on a background of red cells along with inflammatory elements, lymphocytes, histiocytes, and leukocytes. Within the framework of clinical and coelioscopic data, the cytologic pattern assumes definite characteristics permitting the etiologic diagnosis of several gynecologic syndromes. In many cases, it demonstrates premenstrual pelvic congestion and a very marked reaction of peritoneal irritation at the time of ovulation. In other cases, it demonstrates inflammatory phenomena which are clinically inconspicuous or attenuated.

CALLE [29] published the findings of megakaryocytes in the ascites of a patient suffering from megakaryocytic myeloid metaplasia. KORNREICH et al. [111] discussed pericardial disorders in neoplastic diseases. Amongst 1,085 autopsies performed in patients who died from cancer, between 1960 and 1966, 169 cases of pericarditis were found, among which, only 83 were of metastatic origin. Breast cancer was the most frequent cause of pericarditis; leukemia, throat cancer, sarcoma, melanoma, lymphoma, and lung cancer followed in that order of frequency. Only 42 of 169 pericardial lesions were diagnosed before death. The reason for the discrepancy between the clinical and the anatomical observations was the nonspecificity of the clinical manifestations. Uremia as a cause of pericardial effusions was observed in 12 cases by ALFREY et al. [1]. A reliable indicator of cardiac tamponade proved to be a reduction in the cardiac output. If treated correctly, the prognosis of patients with uremic hemopericardium was good. Radiation to the chest may cause pericardial effusion according to case reports by MASLAND et al. [125] and TENG et al. [197]. The background for such effusion is the development of an irreversible pericardial fibrosis [22, 73]. A case of cardiac tamponade secondary to myxedema is described by DAVIS and JACOBSON [49]. The effusion had a distinct 'gold paint' appearance which was due to the suspension of fine cholesterol particles in the pericardial fluid. Several cases of pneumococcal peritonitis as a complication to liver cirrhosis were reported by CAMPBELL [31] and EPSTEIN et al. [60]. CAMPBELL [31] suggested that every patient with pneumococcal peritonitis and ascites should be examined for the presence of occult liver disease. A case of a 54-year-old male who developed a peritoneal mesothelioma following 12 years of pneumoperitoneum was reported by BROWN et al. [23]. Two children with eosinophilic peritonitis containing as much as 90% eosinophils are reported by HUNT et al. [96]. No cause for this condition was detected in either case, but the author believes that the eosinophilia seems to be related to an antigen-antibody reaction. Cytologic specimens obtained by cul-de-sac aspiration of 440 normal women, 35 years and older, were studied by MCGOWAN et al. [132]. The influence of age, the menstrual cycle and the menopause on the differential count of the cells was statistically evaluated. It was found that in the normal woman, a great uniformity existed for every different cell type. The role of the liver in the pathogenesis of ascites was studied by PEARLMAN [156]. He differentiated primary and secondary factors. An increased hepatic intra-vascular pressure supported by hypoalbuminemia is considered the primary factor which is influenced by impaired renal excretion of salt and water, an increased secretion of aldosterone and an excess of antidiuretic hormone in the plasma. VILLA et al. [200] examined 223 cases of ascites with regards to the etiological basis by using the peritoneoscope. They found that simple peritoneoscopy with or without liver biopsy under direct vision proved to be of great value for the diagnosis of some cases of ascites. CEELEN [38] reported a diagnostic accuracy of 92.6% in 160 patients with ascites. Repeat examinations from specimens of the same patients did not change this accuracy. He found that the cell block technique was the most reliable method for examination of the ascites fluid.

In 1969, KERN [105] reported the appearance of psammoma bodies in the culdocentesis fluid of a 54-year-old woman without evidence of ovarian malignancy. This was interpreted by the author as an effect of chronic peritoneal irritation, which may include cancer. RONA et al. [167] published a case of a 56-year-old woman who developed abdominal distention and weight gain 3 years after a mastectomy for cancer of the breast. A paracentesis revealed pure mucus which could be histochemically proven and no cells. Laparotomy revealed a large cystic tumor weighing 6,450 g and containing numerous cystic spaces lined with benign mucus-secreting cells. In one or two areas, a diagnosis of a very well-differentiated cyst adenocarcinoma could be made. Two cases of idiopathic pericardial effusions were reported by WINK and HAGER [206]. They suggested this name because of the absence of an etiological factor for the pericardial effusion. A case of cholesterol pericarditis associated with rheumatoid arthritis was reported by KINDRED et al. [107]. The exudate had the typical 'gold paint' appearance.

A case of massive pericardial effusion due to sarcoidosis of the heart was reported by SHIFF et al. [180]. The diagnosis was proven by biopsy of the characteristic nodules in the pericardium. A massive pleural effusion complicating a case of chronic pancreatitis was reported by MIRIAJANIAN et al. [144]. The pleural fluid contained over 24,000 U of amylase. It was bilateral, hemorrhagic, and accumulated to 17,000 ml in 7 months. CURRAN et al. [44] reported a case of a 43-year-old woman who developed an eosinophilic pleural exudate containing 53% eosinophils. All bacteriological examination was negative, but the patient later developed bronchiectasis. In the past 17 years, the authors have encountered 91 cases of eosinophilic effusions without specific bacteriological findings. Of 25 cases of pericardial effusions in the Sanatoria of North Carolina, 16 were due to tuberculosis, 3 were classified as idiopathic, and 2 were due to malignant tumors. With proper treatment, the prognosis was good, except in the two cases of malignant tumors as reported by SOCHOCKY [184, 185]. WOLFE et al. [208] reported 38 cases of pleural effusions and 60 cases of empyema in infants and children under 16 years of age. All cases responded to adequate drainage and antibiotics. CASTOR and NAYLOR [37] studied the characteristics of normal and malignant human mesothelial cells in vitro. No major difference in the morphological appearance and the growth rate of the cells could be detected. Cytological aspirations from the pelvic peritoneal cavity in pregnant women, postpartum women and normal controls showed a significantly lower number of mesothelial cells and a marked depression of lymphocytes in pregnant women [133]. The latter was interpreted as an increased tolerance to the fetal tissues of the pregnant woman. BOSCARO and CAPRIOGLIO [19] recommend the routine puncture of the Douglas pouch for the detection of ovarian neoplasms. They reported five cases in women between the ages of 22 and 78 in which the cytological examination from the Douglas pouch revealed the presence of an unexpected malignant ovarian neoplasm.

In 1970, BUJA et al. [25] described hemorrhagic pericarditis as a fatal complication in six uremic patients, five of which died from cardiac tamponade. The authors felt that the cause of the hemorrhagic effusion was from pericardial granulation tissue caused by the uremia. SOCHOCKY [184, 185] defines as idiopathic pericardial effusion those cases without evidence of tuberculosis or metastatic cancer. When they persist longer than 8 weeks, he speaks of a chronic idiopathic pericardial effusion. In the history of those patients, an upper respiratory infection with or without pleural effusion usually precedes the onset of the pericardial effusion. This according to the author may point to a viral origin of this condition. LINDSAY et al. [119]

report a case of chronic constrictive pericarditis following a uremic hemo-pericardium and alerts physicians who have patients with uremic pericarditis for this possibility. CARDOZO [34] in 1970 and FLANNERY [67] in 1975 discussed the malignant criteria of pericardial and pleural effusion as well as in ascites. GOVAERTS et al. [80] observed 4 cases of hemorrhagic pericarditis with cardiac tamponade in Bright's disease. Hemorrhagic uremic pericarditis, which is a rare and previously lethal complication of end-stage renal failure, can now be successfully treated by pericardial drainage. Two cases of massive pericardial effusion with cardiac tamponade were observed in cases of myxedema [182]. GAENSLER [75], and GAENSLER and KAPLAN [76] take issue with the term idiopathic for effusions from the chest and the heart. They would like to replace this term with tuberculous effusions even if the tuberculin tests were negative. They also mentioned asbestosis or a connective tissue disease like scleroderma [181], and indicate viral infections of the upper respiratory tract as a possible cause for those pleural effusions. 15 cases of mesothelioma associated with occupational exposure to asbestos were published by MILNE [143] from Victoria, Australia. All cases had a history of exposure to asbestos, especially to crocidolite or blue asbestos. The asbestos bodies were present in the lung tissues of the victims and there was a considerable time span between the exposure and the development of the tumors.

20 cases of diffuse mesothelioma of the pleura were published by ROBERTS [163] with 90% of the cases containing asbestos bodies in the lungs. SCHLIENGER et al. [175] published 39 cases of malignant pleural mesothelioma of which 84% showed metastases. A case of metastatic pheochromocytoma of the pleura was published by TRAUB and ROSENFELD [199] in a patient who died 20 years after the primary tumor had been removed from the adrenal gland. The subcutaneous implantation of cells from a metastatic breast cancer following a needle biopsy of the pleura in a patient with a malignant effusion was reported by JONES [102]. The author feels that such a rare complication does not effect the fate of patients with malignant pleural effusions and it should not be a contraindication to pleural biopsies. BRUNO and OBER [24] report a case of recurrent chylous ascites in an 82-year-old man, which was caused by obliteration of the thoracic duct produced by radiation therapy to the chest. A case of hemorrhagic pleural effusion in pneumonia due to mycoplasm complicated by hemolytic anemia and myocarditis was published by FEIZI et al. [66]. The functional assessment of cytologic findings of pleural cavity effusions was studied by MATZEL [127]. SCHINDLER et al. [174] analyzed 34 patients with chronic pancreatic ascites. They all had a

history of alcoholism, suffered from intermittent abdominal pain and weight loss. The ascitic fluid contained an increased amount of protein and an elevated amylase content of up to 30,000 Somogyi units. Most cases were clinically diagnosed as pancreatic pseudocysts. They all recovered except for four, who died from metabolic disorders. The origin of cardiac effusions was studied by SERVELLE et al. [177]. They showed that transudation of lymph through the myocardial lymphatics was the most prominent cause for pericardial effusions. Resorption of those effusions also occurred through the same lymphatics and to a lesser extent to the pericardial sac. Pleural effusions in cardiac patients is also due to transudation of lymph through the wall of the dilated lymph vessels at the base of the lung. Resorption of the effusion takes place through the diaphragmatic and pleural lymph vessels. Ascites is produced through the transudation of the lymph vessels of the liver. Its resorption takes place through the lymphatics of the small intestines and the lymphatics of the parietal and diaphragmatic peritoneum. STOEBNER et al. [192] studied mesothelial hyperplasia with the electron microscope in patients with metastatic tumors and patients with nonneoplastic lesions. The hyperplastic mesothelial cells were all characterized by great enrichment of the lysosomes which makes them capable of phagocytosis. Their loose connection with each other facilitates their exfoliation. NATELSON et al. [152] wrote about the diagnostic implications of hemorrhagic ascites. It either suggests carcinoma of the ovary or carcinoma of the liver. Occasionally, it is present in a curable disease like tuberculosis of the peritoneum. KIRIANOFF et al. [108] wrote about the diagnostic implications of hemorrhagic ascites. The syndrome of this condition consists of massive ascites, hyperamylasemia of the blood, high amylase levels in the ascitic fluid, and a high protein level of the ascitic fluid. They make a plea for the amylase determination in all ascitic fluid specimens, so that this syndrome will not be overlooked. The spread and localization of intraperitoneal effusions has been studied by MEYERS [141]. The fluid ascends the right external paracolic gutter and enters Morrison's pouch before reaching the right subphrenic space. The phrenicocolic ligament restricts the flow of the fluid between the left paracolic gutter and the parasplenic space. Mesenteric attachments may serve as pathways for the spread of exudate. Pancreatic effusions may pass along the mesentery of the small bowels to the ileocecal region. McGOWAN and DAVIS [134] examined the culdocentesis fluid of 2,677 women between the ages of 16 and 84. They counted 200 cells from each specimen and differentiated six different groups of cells: mesothelial cells, lymphocytes, polymorphonuclear leukocytes, histiocytes, squamous cells, and erythrocytes. By this method, he recognized patterns in human ovulation, in human pregnancy, and in other inflammatory and neoplastic gynecological disorders.

An outstanding contribution to cytology was made by ZACH [213], who discusses the progress of cytology of serous cavities since PAPANICOLAOU. He feels that the study of exfoliated cells in exudates and transudates offers a simple method with better cytological details and good duplication of the results. There are three methods available: the examination of stained preparations, the examination of preparations stained with fluorescent dyes, and the examination of fresh preparations by phase contrast. There exist no single criterium of malignancy, but the diagnosis rests upon a multitude of criteria such as giant cells, giant-sized nuclei with greatly disturbed nucleocytoplasmic ratio, atypical mitoses and abnormal nucleoli. The cells of all serous cavities are similar and cannot be differentiated from each other. In transudates, the endothelial cells appear first in plaque formation, later as single cells with signs of degeneration and monocytic transformation. The greater the variation of the cells the less likely do we deal with cancer. In infarcts, we find mostly endothelial cells and red blood cells. In exudative pleurisy, we find only few mesothelial cells and many lymphocytes. Empyemas show many degenerated leukocytes and bacteria. Eosinophilic effusions have 10–50% eosinophils and are found in allergic conditions and in the so-called idiopathic effusions. Chylous effusions contain many cholesterol crystals and are found in cases of pancreatitis. Malignant effusions usually permit the recognition of malignant cells, but rarely permit the recognition of the tumor type except in cases of melanoma. The histologic examination of the cell block embedded in paraffin greatly compliments the cytological examination.

In 1971, HANCOCK [89] published clinical and hemodynamic observations in 13 patients with pericardial effusions complicated by constriction of the visceral pericardium. The disease was idiopathic in origin in 9 patients and followed radiotherapy in 4 patients. The author emphasized the differentiation of effusion constrictive pericarditis from cardiac tamponade without constriction. The pericardial cellular response during the post-myocardial infarction syndrome was studied by SOLOFF [187] in 1971. The cells in the effusion were composed almost exclusively of leukocytes.

Various types of nonneoplastic pleural effusions have been investigated by a number of authors. BODDINGTON et al. [18] studied 19 cases of pleural effusions complicating rheumatoid arthritis; of those, only 6 showed all characteristics of the disease, while in 9 others the cytological proof was not conclusive. In 12 patients with so-called 'idiopathic' pleural effusion, 21% of the effusions contained asbestos bodies [76]. Those studies point strongly to an intimate connection between the industrial disease of asbestosis and malignant mesothelioma. The problem of pleural effusion and broncho-pleural fistula was studied by MALAVE et al. [122], on 35 patients who suffered from tuberculosis. The establishment of adequate chest drainage followed by operative closure of bronchopleural fistula proved the best therapeutic approach. It is necessary to obtain a negative sputum before surgery. The differential diagnosis of pleural effusions were investigated by a number of authors. KELLEY et al. [103] described atypical cells in the pleural fluid in 8 out of 10 cases of lupus erythematosis. HAIN and ENGEL [88] studied the epidemiology of pleural effusions while BRANDT et al. [21] recommended thoracoscopy for the differential diagnosis of pleural effusions. SAUTER [172] used the methods of cell cultures from cells of pleural effusions or ascites to differentiate benign from malignant cells. OELS et al. [153] reviewed 37 cases of malignant mesothelioma of the pleura, only 10 of which had a history of previous exposure to asbestos. The average survival period of the patient was 18 months. The majority of the cases showed the tubulo-papillary type of mesothelioma. An extensive statistical study of 710 cases of pleural effusions were published by GALY et al. [77, 78].

The anatomy of mesothelial cell effusions was studied by MARSAN et al. [123] on a base of 72 cases. The authors distinguished 3 cytological types: the mesothelial monomorphic type, the mesothelial polymorphic type which is inflammatory, and the mesothelial polymorphic type which is neoplastic. The effect of exposure to asbestos was studied by a number of authors from all parts of the world. WHITWELL and RAWCLIFFE [203] found asbestosis in 60% of their patients with pleural mesothelioma. The interval of the onset of the disease ranged from 13 to 63 years after exposure. Those studies point strongly to an intimate connection between the industrial disease of asbestosis and malignant mesothelioma. Amongst the unusual causes of neoplastic pleural effusions, we wish to mention the case by LABAY and FEINER [113] of malignant pleural endometriosis involving the pelvis, abdomen, and right pleural cavity.

A review of the hormonal abnormalities in the pathogenesis of the edema-ascites syndrome in liver cirrhosis was published by DARNIS [46] and FRANCHI et al. Abdominal sarcoidosis caused a massive ascites in a pregnant woman according to PAPOWITZ and LI [155]. Seven cases of the ascites associated with the Budd-Chiari syndrome were observed by TAKEUCHI et al. [196]. In addition to the ascites, the patients suffered from edema of the legs, although abdominal pain and jaundice were absent. Pancreatic ascites in childhood was discussed by COUPLAND [42]. The changes in the aspects of peritonitis in children was discussed by FOWLER [69]. A tumor diagnosis on the basis of DNA cytophotometry in cells from pleural and ascitic effusions was proposed by FRENI et al. [72]. The Feulgen-DNA content of the nuclei was determined by scanning cytophotometry. The presence of a high incidence of tetraploid cells were used as criteria of malignancy. The technique appears to be too time-consuming to be useful in routine practice. A case of large pseudo-myxoma of the peritoneum originating from a ruptured appendix was published by MOHANDAS et al. [147].

According to DE BACKER et al. [6] in 1972, the sudden appearance of cardiac tamponade is frequently the first manifestation of a metastatic malignant tumor. ZIPF and JOHNSTON [214] examined pericardial effusions of 47 patients for cancer cells. 13 out of 15 patients with tumors involving the pericardium were definitely positive. There were no false-positive reports. The cytologic evaluation of these fluids was helpful for the prognosis and type of therapy. Rheumatoid pericarditis was studied by FRANCO et al. [71]. 93% of the patients showed the rheumatoid factor and in 47% of the patients rheumatoid nodules were present in the skin. The pericardial fluid showed an elevated LDH and γ-globulin level. One third of the patients with rheumatoid pericarditis eventually required pericardiectomy or died of their disease.

The ultrastructure of the pleural mesothelial cell was described by LEGRAND et al. [115, 116], who differentiated 3 different stages of pleural cells: cells at rest, hyperplastic cells, and cells with macrophage activity. MOERTEL [146] commented upon the fact that the mesotheliomas were observed much more frequently in recent years and occurred more commonly in the pleural cavity than in the peritoneal. He confirmed the remarkable frequency of the disease among asbestos miners. In the same year, MURPHY and NG [151] published their extensive study on the determination of the primary tumor site by examination of cancer cells in body fluids. Their studies were based on 87 cases of pleural and peritoneal effusions containing malignant tumor cells from carcinoma of the breast, lung, ovary, endometrium, and stomach. These malignant cells fell into three distinct groups. In group one, the cells occurred in aggregated masses and were uniform. Less than 5% of the cells were vacuolated or multinucleated and less than 5% had macronucleoli. Cells in group two were more isolated, more multinucleated and more vacuolated. Cells of group three contained more than 90% of aggregated cells, more vacuolated cells and more cells with macronucleoli. Based on this classification, the authors were able to determine the primary site of the tumor cells with as much as 96% accuracy. LIGHT et al.

[117] examined 150 pleural effusions by cell counts, protein levels, and LDH levels. Using all three characteristics, it was easier to differentiate exudates from transudates than using either test individually.

WOYKE et al. [209] studied the ultrastructure of alveolar carcinoma of the lungs in the pleural fluid and described characteristic lamellar osmiophilic bodies. BLACK [17] published his results on 436 cases of pleural effusions. Metastatic carcinoma was responsible for 52% of effusions in his series. Congestive heart failure was the second most common cause of pleural effusion which was usually a transudate. JARVI et al. [100] followed a series of 338 cases of pleural effusions with malignant cells in 43 out of 103 cancer patients. The finding of malignant cells always gave a poor prognosis for the patient's life.

SCHMITT et al. [176] recommend intraperitoneal lavage using antibiotics in acute peritonitis. WITTE et al. [207] discussed the effect of portal system obstruction in patients with intra-abdominal neoplasma or peritonitis. A peritoneal fluid with a low protein content signifies a marked impairment of the portal blood flow through the liver. Of unusual tumors, we wish to mention the report of ROBBOY et al. [162] of an ovarian teratoma with glial implants, the report of DE LA MAZA et al. [54] of a case of cervical cylindroma with peritoneal metastases and the report of JACOBS and REYNARD [97] on the content and pathogenesis of chylangioma of the mesentery. Asbestosis was also found present in 3 out of 4 cases of mesothelioma of the peritoneum according to reports of ROBERTS and co-workers [163–165].

Chromosome studies by BENEDICT et al. [11,12] pointed to the presence of long acrocentric marker chromosomes in malignant cells of peritoneal effusions. The same author suggested that chromosomal studies are helpful in confirming the cytological diagnosis, and that positive findings of aneuploidy and marker chromosomes are diagnostic of malignancy. CARDOZO and HARTING [35] investigated the functions of lymphocytes in malignant effusions. They found that they were not much different from normal cells with the exception of increased formation of lymphoblasts. MCGOWAN et al. [132–137] as well as GRAHAM et al. [82] discussed the value of culdocentesis in the examination of peritoneal fluid in women with ovarian tumors. The cells found in the peritoneal specimen resembled closely the tumor cells and it was easy to differentiate benign from malignant ovarian tumors.

In 1973, CARR [36] carried out similar examinations and added bacteriological studies and needle biopsies of the pleura as additional aids to the diagnoosis. His article and that of GRAHAM et al. [81] contain extensive lists of possible causes of pleural effusions. The epidemiology of mesothelioma in Canada was studied by McDONALD *et al.* [130, 131]. The majority of their cases occurred in the pleural cavity. The history of occupational exposure to asbestos was found in 30% of the patients. The same author also reviewed the histology and exfoliative cytology of primary malignant mesothelial tumors. 30% of their cases had had occupational exposure to asbestos. LIGHT *et al.* [117, 118] studied the pleural fluid from 182 patients and found that red blood cell counts of more than 100,000 per mm³ strongly suggested a malignant neoplasm or pulmonary infarction. Lymphocytic predominence suggested tuberculosis or cancer. More than 50% of effusions with inconclusive cytology proved to be due to cancer. When a tumor is suspected, at least 3 separate pleural fluid specimens should be examined.

JAHODA *et al.* [98, 99] studied the computer discrimination of cells in the pleural fluids of eight patients. Their studies indicated that the pleural cell separation by computer evaluation may indeed become a reality in the future. They were encouraged by the fact that the computer could be taught to differentiate between various cellular abnormalities to such an extent that it may offer a better diagnostic accuracy than the human observer. The same authors also published a study on the discrimination of cells by means of computer analysis in peritoneal fluids. They examined normal mesothelial cells, reactive mesothelial cells, and tumor cells from carcinoma of the ovary and esophagus, and found that a satisfactory classification could be obtained for each of the cell types. An unusually high eosinophilia was observed in a case of pleural effusion due to Hodgkin's disease by FARAH *et al.* [64]. FEIZI *et al.* [66] reported a rare case of hemorrhagic pleural effusion with hemolytic anemia due to mycoplasma infection.

ALTEMEIER *et al.* [3] discussed the type and anatomic location of intraabdominal abscesses in 501 cases. They stressed the need for earlier recognition and localization of the abscesses, particularly in patients with pancreatic and retroperitoneal abscesses. McGOWAN *et al.* [135] discussed the value of culdocentesis in the examination of peritoneal fluid in women with ovarian tumors. The cells found in the peritoneal specimen resembled closely the tumor cells and it was easy to differentiate benign from malignant ovarian tumors. ROGOFF *et al.* [166] reported long-term survival in patients with malignant peritoneal mesothelioma treated with irradiation. WEISS [201] studied the accuracy of pleural and peritoneal effusions in his autopsy material. He found 10% false-negative reports and 4% false-positive reports, which produces a total accuracy of his material of 86%.

Pericardial effusions are a frequent complication of scleroderma and many patients suffer from chronic pericardial effusions. In 34 autopsy studies published in 1974, McWhorter and LeRoy [139] recognized pericardial involvement in 62%. Tomb [198] published a cytopathological study on serous fluids in cancer. 114 fluids from 91 patients were examined. In order to obtain the best results, the author emphasized the importance of sending the fluid specimen as soon as possible to the laboratory and upon immediate fixation of the material. Inconclusive cases should be confirmed as soon as possible by histopathological examination of cell blocks or tissue material.

Feizi et al. [66] reported a rare case of hemorrhagic pleural effusion with hemolytic anemia due to mycoplasma infection. Bakalos et al. [8, 9] studied the recognition of malignant cells in pleural effusion by testing them for nonspecific esterases. It is a simple method which can be performed in every laboratory. The same authors also recommend the stain of sudan black-B for the differentiation of monocytic cells from other phagocytic cells. Backman and Pasila [7] discuss the value of pleural biopsy effusions of children. Cailleau et al. [28] isolated four tumor cell lines from pleural effusion from metastatic breast cancer. The cytology of pleural transudates was discussed by Exadaktylos [63]. Pleural lavage lead to the diagnosis of primary carcinoma of the esophagus in a report submitted by Berson [16]. Malignant squamous cells were found in the lavage fluid. The ultrastructure of human mesotheliomas was studied by Davis [48] and Murad [150]. In five cases, the cell surface membranes formed numerous villi of various types and shapes. This feature proved to be very important in the differentiation of mesothelioma cells from cells of lung cancer.

The importance of peritoneal lavage in the diagnosis of acute abdominal problems was studied by Schiller and La Voo [173]. The procedure was without complications and yielded a diagnostic accuracy of 92.5%. Diagnostic lavage appeared to be more accurate than the peritoneal tap in the evaluation of acute abdominal problems [50]. McGowan et al. [132] studied in numerous articles the mesothelial cells of the peritoneal fluid from normal women. They divided the mesothelial cells into three types: small, medium-sized, and hypertrophic types. Each of those may show degenerated forms. Woyke et al. [210] described the ultramicroscopic appearance of hepatoma cells in peritoneal effusion. They possess numerous mitochondria and large golgi complexes. Examination of peritoneal fluid samples in cases of gynecological patients has lead to inconclusive reports according to Maathuis [121], but he recommends to continue the physical and biochemical examination of the peritoneal fluid in addition to the morphological studies. Two cases of malignant peritoneal mesothelioma were diagnosed before death by Eslami and Lutcher [61]. Peritoneoscopy was very helpful in the antemortem diagnosis of this rare type of neoplasm. Pancreatic ascites was studied by DONO-WITZ et al. [57]. This syndrome refers to the accumulation of massive amounts of ascites in cases of chronic or acute pancreatitis. The nature of the ascites is not well understood and no diagnostic laboratory tests exist. In about 85% of the patients, a pseudocyst of the pancreas or a ruptured pancreatic duct may be found at surgery, but in many cases the pathogenesis of the ascites during the course of pancreatic disease is not clear. Five patients with chronic kidney disease undergoing intermittent hemodialysis developed massive ascites according to CRAIG et al. [43]. The protein content and LDH level of the ascitic fluid were high, which differentiated this type of ascites from ascites in liver cirrhosis. Other causes for the ascites in those patients treated by hemodialysis were ruled out by clinical data, peritoneal fluid studies, and peritoneoscopy.

In 1975, LOIRE et al. [120] recommend pericardial biopsies through a left lateral thoracotomy as an aid for the diagnosis of effusions of the pericardium. By this method, they were able to recognize radiation pericarditis and cases of idiopathic pericarditis. Draining of a pericardial effusion was recommended as successful treatment in cases of pericardial effusions in patients with malignant disease [67]. Radiation therapy lead to a 60% improvement in patients with pericardial metastases according to CHAM et al. [39].

A comparison between the efficacy of the needle biopsy and pleural fluid cytopathology was made by SALYER et al. [169] in 271 patients. Using both methods, the authors achieved a diagnostic accuracy of 90%. A recent editorial of British Medical Journal [59] discusses the difference of pleural exudates and pleural transudates. They recommend cultures of pleural fluid and pleural biopsy as an additional aid in the differential diagnosis. HOFF and LIVOLSI [93] discuss the reliability of needle biopsy of the parietal pleura in 272 cases. Only two false-positive diagnoses were made and a negative pleural biopsy should not give the clinician a false sense of security. SHARMA and GORDONSON [179] reported six cases of pleural effusion in sarcoidosis. Their diagnosis was based upon the histological evidence of a pleural biopsy demonstrating noncaseating granulomas of the disease.

Tissue culture techniques are helpful in the diagnosis of metastatic carcinoma in peritoneal fluids according to MONIF and DALY [148]. In 15 out of 18 cases, small islands of growing carcinoma cells could be observed after 6 days of culture. The value of peritoneal lavage in the diagnosis of acute peritonitis was investigated by EVANS et al. [62]. In 11 out of 34 patients, the lavage corrected a false clinical diagnosis. Another appraisal of the diagnostic value of peritoneal lavage in acute abdominal pain was made by BARBEE and GILSDORF [10]. They obtained 28% false-negative and 8% false-positive results and 64% correct results with lavage which was most accurate in cases of appendicitis, colonic disease, and intra-abdominal bleeding. They recommend that the procedure may be helpful in a difficult diagnostic problem, but should not be employed routinely. A case of massive ascites in a patient with plasmacytoma of the intestine was reported by HIGBY and OHNUMA [91]. There was no skeletal or bone marrow involvement.

Transmission and scanning electron microscopic studies of cancer cells and benign cells from effusion were reported by DOMAGALA and WOYKE [56]. Fundamental differences were described in benign and malignant cells and the authors recommend this approach for the diagnosis of difficult cases and for investigative purposes. SPRIGGS [190] investigated the various techniques for removing red blood cells from hemorrhagic serous fluids. He recommends the use of 25% Hypaque. By this method, the white cells can be found in the upper layer of the centrifugate, the red cells at the bottom and the malignant cells and mesothelial cells between both layers. DEKKER et al. [53] described the occurrence of sickle cells in a 20-year-old patient in sickle cell crisis. The cells probably originated from a pulmonary infarct.

III. The Normal Cytology and Histology of the Serous Membranes

The body cavities form from the embryonic coelom and the first and largest cavity is the pericardial cavity. Later, by the formation of the pleuro-pericardial fold and the pleuroperitoneal fold, the other body cavities are formed which, in the fully developed human being, represent the serous cavities containing the heart, both lungs, and the visceral organs. In the adult, those serous membranes – the peritoneum, pleura, and pericardium – are thin layers of connective tissue covered by a single layer of mesothelium. With the exception of the peritoneal cavity, the other cavities are slit-like spaces which contain a minimum amount of fluid secreted by the meso-thelial cells. They contain a very small amount of fluid which serves as lubricating material to facilitate movements of the organs contained in those cavities. The histology of the serous membranes can be divided into the serosal layer and the subserosal layers. The serosal layer is covered by a single layer of mesothelial cells. Those are cuboidal cells which are highly vulnerable and which possess a fine brush border. Electron microscopically, those cells possess a large nucleus, few villi, and a demonstrable brush border. They are only very loosely connected with the subserous layer from which they easily exfoliate into the body cavities. The subserous layer consists of strands of nonspecific fibrous tissue without any specific functional qualifica-tions, such as elastic fibers or polysaccharides. They contain a small net-work of capillaries, lymphatics, and nerve fibers. They are easily invaded by histiocytes and a variety of inflammatory cells [74].

The most important cells of the serous membranes are the mesothelial cells. They differ in size and small cells, medium-sized cells, and large or hyperplastic forms have been described. They appear either singly or in clusters or sheets. They are usually round or oval and measure between 10 and 20 μm in diameter. They possess a homogenous basophilic cytoplasm without any vacuoles or phagocytosed material. The nucleus is round and centrally located. It has a prominent nuclear membrane, a fine chromatin network, and occasionally one or two small nucleoli. Binucleated and multi-nucleated forms are not uncommon. In cell blocks, they often form pseudo-

glandular structures or rosettes which may represent an artifact of centri-
fugation. At the slightest irritation, the mesothelial cells undergo characteristic changes which are called active forms or reactive forms by various
observers [74]. The cells enlarge in size, the nucleus assumes a sharply demar-
cated nuclear membrane and multinucleated forms appear. Degenerative
forms amongst mesothelial cells are quite common and can be found in the
normal cells as well as in the reactive forms. The first evidence of degeneration
is the appearance of fatty or hydropic vacuolization. The nuclei become pale
and the chromatin becomes coarse. The cytoplasm loses its homogenous
appearance and becomes quite vacuolated. In some instances, those vacuoles
are so large that they push the nucleus against the cell wall so that the cells
assume the shape of signet ring cells. While normal mesothelial cells usually
participate very little in the process of phagocytosis, degenerated mesothelial
cells will permit the entrance of foreign material in the form of granules or
droplets. Thus, those cells resemble macrophages which belong to the series
of histiocytes. Sometimes, one observes cells which SPRIGGS [190] has given
the nickname of 'pseudoplasma cells' which represent a degenerated mesothelial cell with a deeply basophilic cytoplasm. Abnormal mesothelial cells
often make their differentiation from cells of metastatic tumors exceedingly
difficult. Mesothelial cells in cases of scleroderma or amyloidosis belong with
this group [112]. Radiation or chemotherapy may produce substantial
changes in the mesothelial cells with hyperchromasia of the nuclei and
bizarre cell forms. We shall discuss those changes in detail in another
paragraph [87].

Other cells observed in normal serous membranes have migrated through
capillaries and lymphatics. SPRIGGS and BODDINGTON [189] mention lymphocytes, plasma cells, neutrophil leukocytes, eosinophil leukocytes, mast cells,
red blood cells, and occasionally megakaryocytes. SOOST [188] mentions the
macrophages which are cells capable of phagocytosis and which originate
from the histiocytes. However, degenerated mesothelial cells may resemble
those cells since they also have vacuoles and can exhibit the capability of
phagocytosis. Attempts to separate those two cell types by special stains
have thus far not been satisfactory. A variant of plasma cells which are
occasionally found amongst the cells of the serous membranes are the so-
called Mott cells, which have granules which stain strongly with PAS even
after digestion with saliva.

IV. The Cytology of Transudates

Transudates represent the accumulation of fluids in the serous cavities, mostly due to increased venous pressure and other circulatory disturbances. TAKAHASHI [195] quotes cardiac insufficiency, hepatic cirrhosis, renal insufficiency, and hypoproteinemia as the principal causes of a transudate. It is characterized by a specific gravity of less than 1,015 and a protein content of less than 2 g%. It usually has a clear color with a faint yellowish tinge. The principal cellular elements of a transudate are the mesothelial cells. Those cells are cuboidal in shape and desquamate easily into the fluid of the transudate, where they accumulate either as single cells or in clusters, sheets, or placards. Less common is the formation of papillary masses or rosettes. Figure 1 portrays normal and reactive mesothelial cells in proteinaceous fluid. Under the reactive mesothelial cells, we understand cells with an enlarged nucleus and prominent nucleoli. Figure 2 represents fluid from a patient with congestive heart failure depicting normal mesothelial cells in a rosette arrangement. Figure 3 represents reactive mesothelial cells from pericardial fluid in a case of congestive heart failure. Again note the enlarged nucleus and the homogenous cytoplasm. Figure 4 represents mesothelial cells from the ascites of a patient with congestive heart failure. Here mesothelial cells show evidence of degeneration with pyknotic nuclei and vacuolization of the cytoplasm (fig. 5).

Figure 5 shows transudate in a case of chronic congestive heart failure. The cell block preparation shows many active forms of mesothelial cells together with a few lymphocytes and histiocytes. Figure 6 represents again cells in pleural fluid in a case of congestive heart failure with many atypical and partly degenerated mesothelial cells. Figure 7 shows clumps of degenerated mesothelial cells with pyknotic nuclei and dark cytoplasm mixed with histiocytes. The cells show vacuoles and large pale nuclei with prominent nucleoli. Figure 8 is a specimen taken from the pleural fluid in a case of congestive heart failure. It shows numerous reactive mesothelial cells with pyknotic nuclei and vacuolization of the cytoplasm. Figure 9 represents the pleural fluid in a case of chronic congestive heart failure. It shows lymphocytes and many reactive and degenerated mesothelial cells. Figure 10 represents another pleural fluid in a case of congestive heart failure and shows a cluster of red blood cells, inflammatory cells, and reactive mesothelial cells. Figure 11 shows for comparison a clump of cells from a case of poorly differentiated adenocarcinoma, probably arising from the breast. Figure 12 represents a cell block preparation of pleural fluid from a case of congestive heart failure, showing pseudorosettes of mesothelial cells mixed with numerous histiocytes.

The ascites formed in cases of liver cirrhosis is characterized by the presence of many degenerated mesothelial cells [161]. They show usually eccentric nuclei and many vacuoles in the cytoplasm. They resemble in many respects histiocytes with the exception that they lack phagocytosed material (fig. 13). They usually show fine granules which stain red with the PAS stain. Although this stain in our experience does not represent a foolproof characteristic sign of mesothelial cells, it is quite helpful in differentiating them from histiocytes.

Since it is quite often impossible to differentiate between degenerated forms of mesothelial cells and active histiocytes, it has become the custom to consider both cell types as macrophages. Those cells have as their only common characteristic the ability of phagocytosis. They may originate from degenerated mesothelial cells, from histiocytes, from plasma cells, lymphocytes, and monocytes.

Figure 14 represents a pseudogland formed by reactive mesothelial cells, in the ascites from a case of liver cirrhosis. The lack of oxygen in the ascites fluid of this disease produced a characteristic pyknosis of the mesothelial cells which possess a deeply cyanophilic plasma which is often hard to differentiate from the pyknotic and deeply cyanophilic nucleus (fig. 15). The nuclear pyknosis is sometimes so pronounced that it is difficult to differentiate the cell from a malignant hepatoma cell as exemplified in figure 16.

Other transudates are not caused by failure of the circulation, but by changes in the composition of the transudate which facilitates its permeability through the walls of the capillaries and lymph vessels. Chronic kidney disease is a typical example for the formation of a transudate of this type. Figure 17 shows a pleural effusion with mesothelial cells and lymphocytes in a case of chronic glomerulonephritis. Note the edematous condition of the lymphocytes. The same can be observed in figure 18, which represents ascites in a patient with chronic liver and kidney disease. Many mesothelial cells show vacuoles and cannot be differentiated from the signet ring-shaped edematous histiocytes. Figure 19 represents transudate of a pleural fluid in a case of chronic kidney disease. It contains numerous reactive mesothelial cells of different sizes and shapes. Figure 20 represents a pleural fluid transudate from a similar case. It is rich with large reactive mesothelial cells.

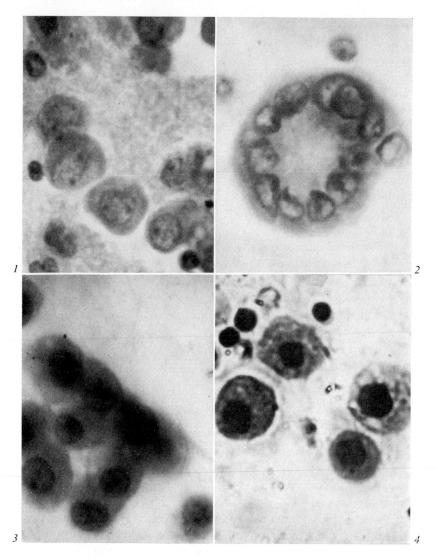

Fig. 1. Congestive heart failure. Pleural fluid. Normal and reactive mesothelial cells and a few histiocytes. Papanicolaou. × 880.

  Fig. 2. Congestive heart failure. Pleural fluid. Rosette of mesothelial cells. Cell block preparation. × 880.

  Fig. 3. Congestive heart failure. Pericardial fluid. Reactive mesothelial cells in small clumps. Papanicolaou. × 1,100.

  Fig. 4. Chronic effusion in patient with heart failure. Ascites. Pyknotic mesothelial cells with signs of anoxia and degeneration. Papanicolaou. × 880.

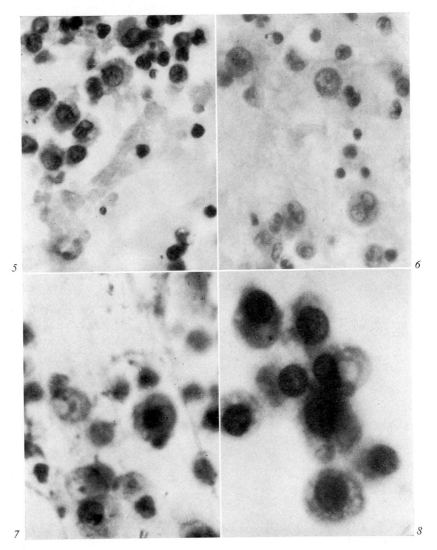

Fig. 5. Chronic congestive heart failure. Pleural fluid. Transudate with reactive mesothelial cells and leukocytes. Cell block preparation. × 700.

Fig. 6. Congestive heart failure. Pleural fluid. Many reactive atypical mesothelial
cells and a few lymphocytes. Papanicolaou. × 550.

Fig. 7. Congestive heart failure. Pleural fluid. Clumps of mesothelial cells and
histiocytes. Papanicolaou. × 1,100.

Fig. 8. Congestive heart failure. Pleural fluid. Reactive mesothelial cells. Papanicolaou. × 880.

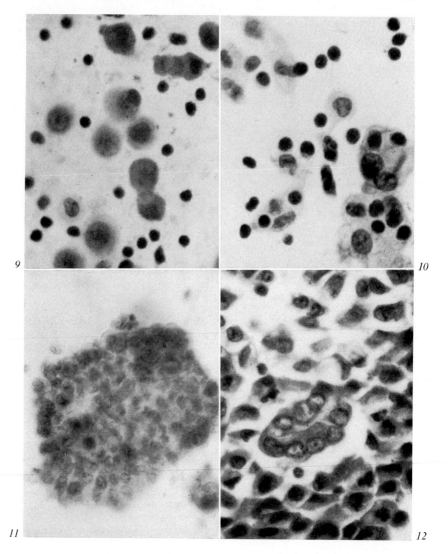

Fig. 9. Chronic congestive heart failure. Pleural fluid. Reactive mesothelial cells and a few lymphocytes. Papanicolaou. × 550.

Fig. 10. Congestive heart failure. Pleural fluid. Clusters of reactive mesothelial cells, inflammatory cells, and red blood cells. Papanicolaou. × 550.

Fig. 11. Congestive heart failure. Pleural fluid. Many clones of poorly differentiated adenocarcinoma cells probably arising from the breast. Papanicolaou. ×265.

Fig. 12. Chronic congestive heart failure. Pleural fluid. Mesothelial cells forming a pseudorosette and histiocytes. Cell block preparation. × 550.

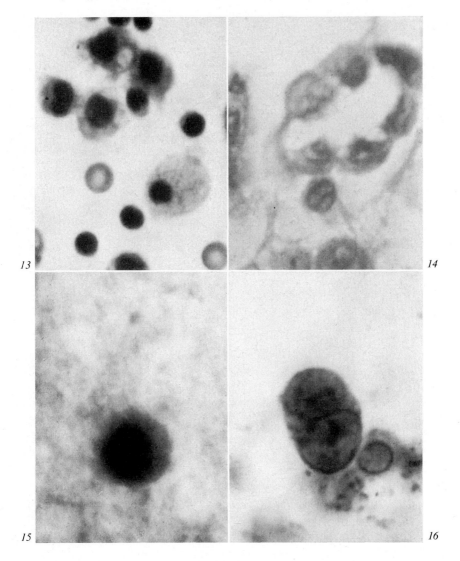

Fig. 13. Cirrhosis of the liver. Ascites. Many mesothelial cells and signet ring-shaped edematous histiocytes. Papanicolaou. × 1,100.

Fig. 14. Cirrhosis of the liver. Ascites. Pseudogland formed by reactive mesothelial cells. Cell block preparation. × 1,100.

Fig. 15. Cirrhosis of the liver. Pleural fluid. Mesothelial cell as evidence of anoxia. Papanicolaou. × 1,750.

Fig. 16. Chronic liver disease. Ascites. Single large malignant cell resembling a hepatoma cell with two nuclei and bizarre nucleoli. Papanicolaou. × 550.

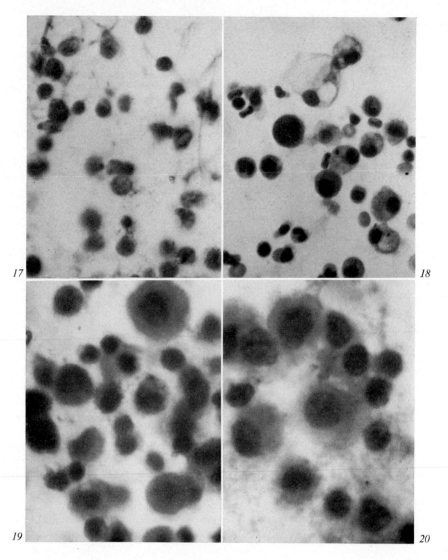

Fig. 17. Renal failure. Pleural fluid. Lymphocytic transudate with endothelial cells and lymphocytes. Papanicolaou. × 550.

Fig. 18. Chronic liver and kidney disease. Ascites. Many mesothelial cells and signet ring-shaped edematous histiocytes. Papanicolaou. × 550.

Fig. 19. Chronic kidney disease. Pleural fluid. Atypical reactive mesothelial cells varying in size. Papanicolaou. × 880.

Fig. 20. Chronic kidney disease. Pleural fluid. Many reactive benign mesothelial cells. Papanicolaou. × 1,100.

V. The Cytology of Inflammatory Exudates

Exudates can be divided into two large groups: those produced by benign inflammatory processes and those produced by primary or metastatic tumors to the pleural cavities. Acute inflammatory processes can reach the serous cavities either directly, like in trauma, or by an infection spreading from an organ situated in the serous cavity, like the lung, or by the spread of an infection to the serous cavity by lymph vessels or blood vessels. Infection of any of the serous cavities will lead to an exudate which varies according to the infectious agent, its virulence, the reactivity of the host, and the duration of time; thus, we speak of acute inflammatory processes and chronic inflammatory processes. A very common instance of an acute inflammatory process is seen in figure 21, which represents a smear from a case of purulent pleurisy or empyema, developing from a severe pneumonia. A rich proteinaceous background with many thousands of leukocytes is visible. They are all neutrophil leukocytes and they all show evidence of severe cell injury. Some of them are swollen, others are shrunken, and many of the nuclei are necrotic. This picture was produced by infection of the pleural cavity with a very virulent pyogenic organism. Figure 22 shows also an acute inflammatory exudate taken from a case of myocarditis with pericardial effusion. It is evident that the white blood cells are better preserved and that there are quite a few reactive mesothelial cells in the exudate. The difference between both pictures lies in the severity of the infectious agent, a highly virulent pneumococcus in figure 21 and a virus infection in figure 22. Inflammatory exudates will remain in the serous cavities as long as the etiological agent is active and in those cases we speak of acute pericarditis, acute pleurisy or empyema, and acute peritonitis. Once the forces of defense have killed the infectious organisms, the exudates are either absorbed or try to escape by a natural route such as perforation into a bronchus. The cells in those exudates undergo severe degenerative changes and also changes in the cell population. The leukocytes are replaced by reactive mesothelial cells and lymphocytes (fig. 23). Figure 24 shows a similar case of chronic pleural effusion with reactive mesothelial cells and lymphocytes.

Figure 25 was taken from the pleural fluid of a very acute lobar pneumonia. The infectious process had not yet invaded the pleural cavity, but the mesothelial cells show severe degenerative changes reflecting the toxicity of the process. They are all in a highly reactive state with many degenerative vacuoles. Figure 26 has been taken from a hemorrhagic fluid caused by a pulmonary infarction. We see clumps of reactive mesothelial cells, inflammatory cells and many red blood cells. Sometimes these effusions carry a high percentage of eosinophilia. Figure 27 represents s cell block taken from a chronic pleural effusion, probably of postpneumonia origin. We see groups of histiocytes and reactive mesothelial cells. Figure 28 has been taken from an ascites in a case of chronic peritonitis. It represents a cell block showing a mixture of reactive mesothelial cells, many histiocytes, and a few leukocytes. Figure 29 represents a chronic pleural effusion from a case of chronic pulmonary disease with bronchiectasia and chronic pneumonia. The majority of the cells are large reactive mesothelial cells with a few scattered lymphocytes. Figure 30 has been taken from a case of confluent bronchopneumonia with numerous degenerated mesothelial cells and some large histiocytes containing numerous cytoplasmic vacuoles. Figure 31 also has been taken from a case of confluent bronchopneumonia. The process obviously had not yet spread to the pleural cavity, but we see numerous large and active mesothelial cells as evidence of pleural irritation. There are also few histiocytes and neutrophil leukocytes. Figure 32 is also a pleural effusion from a case of bronchopneumonia which shows a large giant cell probably formed by reactive mesothelial cells. This giant cell must not be mistaken for the Langhans giant cells that we find in tuberculosis, but resembles more a foreign body giant cell.

Figure 33 is a pleural effusion from a case of pulmonary tuberculosis. The majority of the cells are small lymphocytes which appear either singly or in small group. There are few degenerated mesothelial cells present. Figure 34 is a similar case, tuberculosis studied under high magnification. It contains mostly lymphocytes and a large percentage of eosinophilic leukocytes. The latter finding is unusual as has been emphasized by SPRIGGS and BODDINGTON [189]. Our case contained approximately 90%. Figure 35 represents a case of tuberculous pleurisy which was proven by pleural biopsy. The exudate is rich in lymphocytes and contains few reactive mesothelial cells. Figure 36 represents a case of chronic pulmonary tuberculosis. Again we see a majority of lymphocytes in the exudate in addition to many atypical mesothelial cells and histiocytes. Figure 37 represents pericardial fluid from a case of rheumatic heart disease. Small mesothelial cells are gathered in rosette formation amongst a background of fibrous material and red blood cells. Figure 38 represents the pleural fluid from a patient with rheumatoid arthritis. The exudate consists of structureless proteinaceous material with one large binucleated spindle cell which is characteristic for the disease. Figure 39 shows a pleural biopsy of the same case. Note the pallisade-like arrangement of the identical spindle cells as we have demonstrated in figure 38 on a background of fibrosis and necrosis.

Figure 40 is a case of idiopathic eosinophilic effusions. Eosinophilia is quite a frequent observation in effusion and, according to Koss [112], can be associated with many pathological states, such as hypersensitivity, pulmonary infarction, malignant neoplasm, and allergic diseases. On occasion, no associated disease is observed and surgical exploration reveals no distinctive pathological features. BOWER [20] calls eosinophilic pleural effusion a condition with multiple causes and also stresses the fact that in a significant number of cases the cause of the effusion is unexplained. It may or it may not be accompanied by blood eosinophilia. Unless associated with tumor, the prognosis of eosinophilic pleural effusion is good. Its duration usually does not extend more than a few weeks. The fluid in those conditions is usually blood-tinged and is usually sterile. The number of eosinophils may reach more than 90%. Koss [112] names this ill-defined entity 'idiopathic eosinophilic pleural effusion' and several authors attribute the eosinophilia to hypersensitivity secondary to nonspecific respiratory infection.

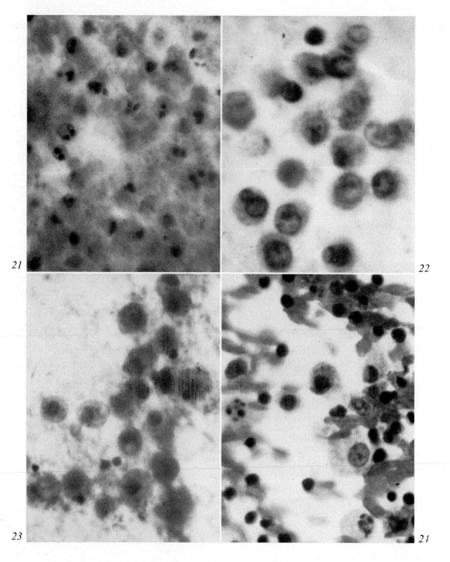

Fig. 21. Acute pleuracy. Pleural fluid. Exudate consisting almost entirely of necrotic leukocytes (empyema). Papanicolaou. × 550.

Fig. 22. Acute myocarditis. Pericardial fluid. Blood mixed with reactive mesothelial cells and histiocytes. Papanicolaou. × 550.

Fig. 23. Chronic pneumonitis. Pleural fluid. Many reactive mesothelial cells, a few lymphocytes, and red blood cells. No tumor cells. Papanicolaou. × 1,100.

Fig. 24. Chronic pulmonary disease. Pleural fluid. Reactive mesothelial cells in a background of blood. No malignant cells. Papanicolaou. × 550.

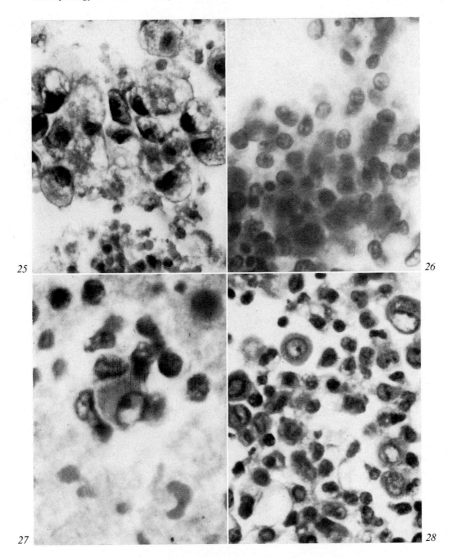

Fig. 25. Lobar pneumonia. Pleural fluid. Reactive mesothelial cells with foamy cytoplasm without phagocytosis. Papanicolaou. × 550.

Fig. 26. Pulmonary infarction. Pleural fluid. Clump of reactive mesothelial cells and inflammatory cells. Papanicolaou. × 550.

Fig. 27. Chronic pleuracy. Pleural fluid. Groups of histiocytes and mesothelial cells. Cell block preparation. × 1,100.

Fig. 28. Chronic peritonitis. Ascites. Mesothelial cells, histiocytes, and mild inflammation in a background of blood. Cell block preparation. × 550.

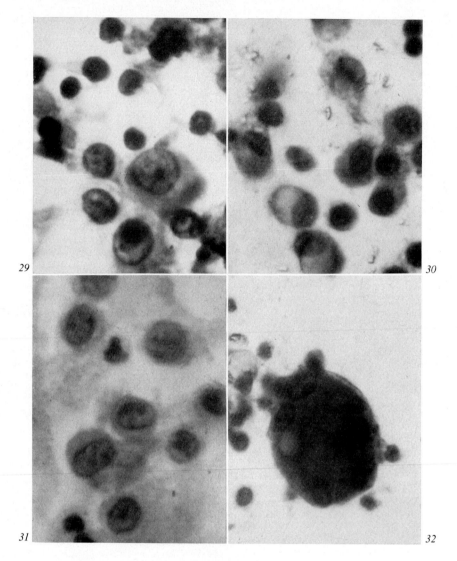

Fig. 29. Chronic pulmonary disease. Pleural fluid. Reactive mesothelial cells in a background of blood. No malignant cells. Papanicolaou. × 1,100.

Fig. 30. Pneumonia. Pleural fluid. Edematous histiocytes and reactive mesothelial cells with a few lymphocytes. Papanicolaou. × 880.

Fig. 31. Bronchial pneumonia. Pleural fluid. Mesothelial cells, histiocytes, and inflammatory cells. Cell block preparation. × 1,100.

Fig. 32. Bronchial pneumonia. Pleural fluid. Giant cell formed by reactive mesothelial cells. Papanicolaou. × 550.

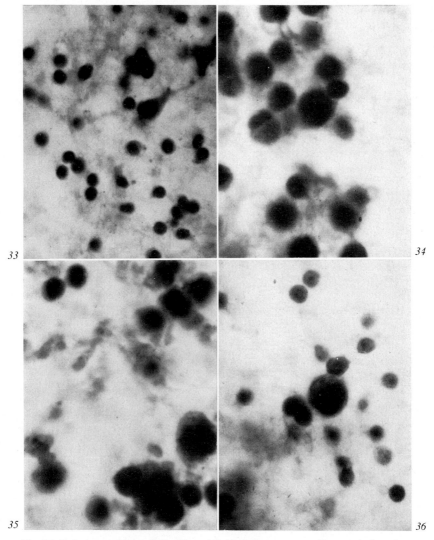

Fig. 33. Pulmonary tuberculosis. Pleural fluid. Inflammatory cells, mostly lymphocytes with a few mesothelial cells. Papanicolaou. × 550.

Fig. 34. Pulmonary tuberculosis. Pleural fluid. Blood, very rare mesothelial cells and moderate inflammation with approximately 90% eosinophils. Papanicolaou. × 1,100.

Fig. 35. Tuberculous pleuracy. Pleural fluid. Mesothelial cells and many lymphocytes. No malignant cells. Papanicolaou. × 1,100.

Fig. 36. Chronic tuberculosis. Pleural fluid. Many atypical mesothelial cells and histiocytes consistent with changes associated with tuberculous effusions. Papanicolaou. × 550.

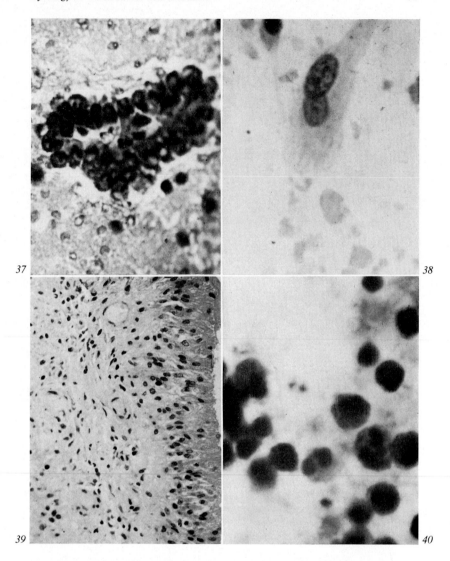

Fig. 37. Rheumatic heart disease. Pericardial fluid. Rosette formation of normal pericardial mesothelial cells. Papanicolaou. × 550.

Fig. 38. Rheumatoid disease. Pleural fluid. Large pale spindle cells in proteinaceous material. Papanicolaou. × 550.

Fig. 39. Rheumatoid disease. Pleural biopsy. Rheumatoid nodule. Papanicolaou. × 271.

Fig. 40. Eosinophilic effusion. Pleural fluid. 66% eosinophils in sediment. Papanicolaou. × 1,100.

VI. The Cytology of Malignant Exudates

By a malignant exudate, we understand a fluid accumulating in one or more of the serous cavities which is caused by a malignant tumor. This malignant tumor may be a primary tumor of the serous cavity which is called a mesothelioma or, more often, it may be a metastasis to the serous cavity from a malignant tumor of any other organ in the human body. The response to a primary or metastatic tumor is the formation of an effusion which has all the chemical and gross characteristics of an exudate with its high protein content, its high specific gravity and the presence or absence of a hemorrhagic component. In addition to all those characteristics, and in contrast to benign exudates which we discussed in the previous chapter, this type of exudate also contains malignant cells, which can be found amongst the other cells of the exudate. Since the fluid from serous exudates usually represents a good nutrient media, it usually enhances the growth of cells including that of malignant cells. For this reason, it is usually not difficult to recognize malignant cells in large number together with the growth of mesothelial cells. The difficult problem of the cytopathologist is to recognize the original site of the malignant cells so he can tell the clinician their origin. In undifferentiated tumors, this may be an impossible task, but in tumors where the cells show some evidence of cytoplasmic or nuclear differentiation, it may be possible to at least suggest one or two primary sites for the metastasizing cancer. We have in our collection in addition to primary tumors of the serous cavities, metastases from 19 different malignant tumors which we could identify according to their origins. The majority of them belong to the group of carcinomas. We shall point out in the following pages some of the features by which primary or metastatic tumors of the serous cavities can be recognized. We shall begin with the group of primary tumors of the serous cavities or mesotheliomas.

Mesotheliomas of the Serous Cavities

Mesotheliomas may originate in the pericardial, pleural, and peritoneal cavities. They used to be considered very rare tumors, but recent surveys have shown a steady increase of mesotheliomas, particularly those arising in the pleural cavity [45, 90]. They either show a nodular spread or form thick plaque-like tumor masses which interfere with the mobility of the heart or the lungs. They always produce an exudate with abundant numbers of tumor cells [159]. They extend rapidly throughout the serous cavities and produce multiple nodular implants or fill the cavity with solid masses of fleshy neoplastic tissue. Figures 41 and 42 show malignant mesothelial cells characteristic of a carcinomatous mesothelioma. The cells are single, irregular in size, and possess large pyknotic nuclei. Figure 43 shows a morula-like clump of malignant mesothelioma cells while figure 44 shows an isolated giant-sized cell from a case of malignant mesothelioma. Note the central position of the nucleus, the multiple nucleoli, and the opaque appearance of the cytoplasm. It is cells of this type which make the cytological diagnosis of mesothelioma possible, since in many cases the differentiation of malignant from benign mesothelial cells is extremely difficult [112].

Figure 45 represents cells from a mesothelioma of the peritoneal cavity. Note the similarity of the large malignant mesothelial cells and the great variation in the size and shape of the cells. Many of the cells show degenerative vacuolization. Figure 46 is a cell block preparation of the same case. There are many groups of malignant mesothelial cells characterized by irregular size and multinucleation. The picture contains one group of cells resembling a papillary cluster. Figures 47 and 48 are specimens taken from a mesothelioma of the pericardial cavity. The cells are much more regular and many resemble reactive mesothelial cells. One group of cells in figure 47 shows distinctly cannibalism, while one cell cluster in figure 48 represents undoubtedly malignant mesothelial cells.

Numerous reports have appeared in the last 10 years which link malignant mesotheliomas with the asbestos industry [41]. Asbestos miners and persons handling merchandise made from asbestos develop mesotheliomas much more frequently than the population at large. Most of those mesotheliomas develop in the pleural cavities and the time interval between exposure to asbestos and the development of the mesothelioma may run from 10 to 30 years. More data pointing to the causal relationship between asbestos and mesothelioma may be found in our survey of the literature in chapter II.

Carcinoma of the Breast

The two most common malignant tumors which metastasize to the serous cavities are undoubtedly the carcinomas of the breast and of the lungs.

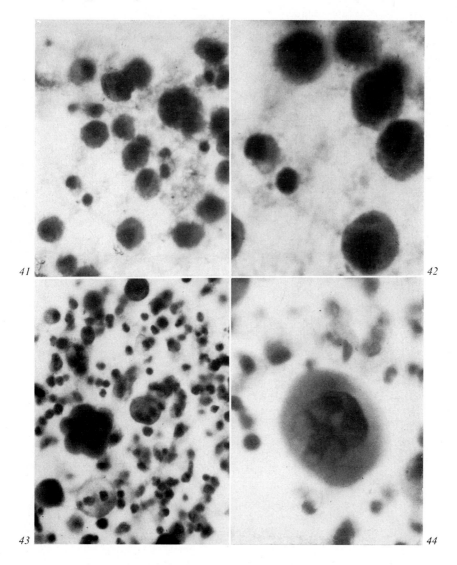

Fig. 41. Persistent pleural effusion. Pleural fluid. Malignant mesothelial cells characteristic of mesothelioma of the pleura. Papanicolaou. × 550.

Fig. 42. Persistent pleural effusion. Pleural fluid. Malignant mesothelial cells characteristic of mesothelioma of the pleura. Papanicolaou. × 1,100.

Fig. 43. Mesothelioma. Pleural fluid. Clumps of epithelial-like cells, mixed vesicular cells and inflammatory cells. Papanicolaou. × 550.

Fig. 44. Mesothelioma. Pleural fluid. Gigantic tumor cell with bizarre nucleus and large nucleolus. Papanicolaou. × 1,100.

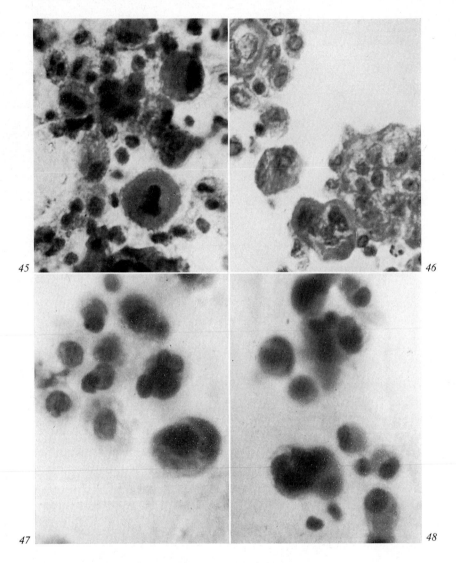

Fig. 45. Mesothelioma. Peritoneal fluid. Many large and small malignant mesothelial cells. Papanicolaou. × 550.

Fig. 46. Mesothelioma. Peritoneal fluid. Many groups of atypical mesothelial cells
characterized by many malignant multinucleated cells. Cell block preparation. × 550.

Fig. 47. Mesothelioma. Pericardial fluid. Malignant mesothelial cells with transition
to reactive mesothelial cells. Papanicolaou. × 700.

Fig. 48. Mesothelioma. Pericardial fluid. Malignant mesothelial cells with transition
to reactive mesothelial cells. Papanicolaou. × 700.

Both metastasize more frequently to the pleural cavity than to the other serous cavities. This may be easily explained by the rich network of lymphatics which link the breast glands and the lungs to the pleura. Nevertheless, the other serous cavities are also frequently involved. Histopathologists differentiate twelve different types of carcinomas of the breast, all of which can metastasize to the serous cavities. The most common are the small cell carcinomas of the breast, or scirrhous carcinomas. They appear in the pleural fluid in the form of single cells or small rows of three to four cells. They possess a large nucleus with a distinct nucleolus and a cytoplasm which usually stains pale blue or pink. Figures 49 and 50 are examples of single undifferentiated carcinoma cells from the breast in the pleural fluid. Were it not for their large nuclei, they could easily be taken for mesothelial cells. Figure 51 shows a small row of four malignant cells separated by a slightly vacuolated cytoplasm. Figure 52 shows a group of three small malignant cells arranged in a typical single row which Koss [112] has compared with a soldiers' file. Scirrhous carcinomas of the breast always carry a large fibroblastic component in the primary tumor. Metastatic scirrhous carcinomas may lose this fibroblastic component and appear as a small cell undifferentiated carcinoma.

Figure 53 shows a mixture of undifferentiated malignant cells from a carcinoma of the breast with very edematous mesothelial cells. Notice the remarkable difference in the nuclear-cytoplasmic ratio of both cell types. It probably is derived from the mucous-secreting cells of the larger breast ducts. The cells stain positive with PAS stain as well as mucicarmin. Figure 54 represents a group of malignant breast cells taken from the pericardial fluid. It shows a cluster of mucous-containing adenocarcinomatous cells. Figure 55 shows a large group of malignant cells from an adenocarcinoma of the breast metastasizing to the pericardium. The cells contain a large amount of mucous and mucous can also be found outside the cell. We are dealing here with a rare colloid carcinoma of the breast. Figure 56 represents a highly atypical adenocarcinoma of the breast which has metastasized to the pleural cavity. We find some very large cells with a high content of mucous and clusters of smaller cells which are also of the adenocarcinomatous type. We feel that this tumor had also developed from the mucous-producing glands ot the larger breast ducts. Figure 57 represents a small cluster of small adenocarcinoma cells containing little mucous. It probably represents a lobular type of cancer of the breast, which has metastasized to the pleura. Figure 58 shows cells resembling figure 57 with the exception that the cells contain larger mucous droplets. We feel that this represents a transitional stage from lobular to

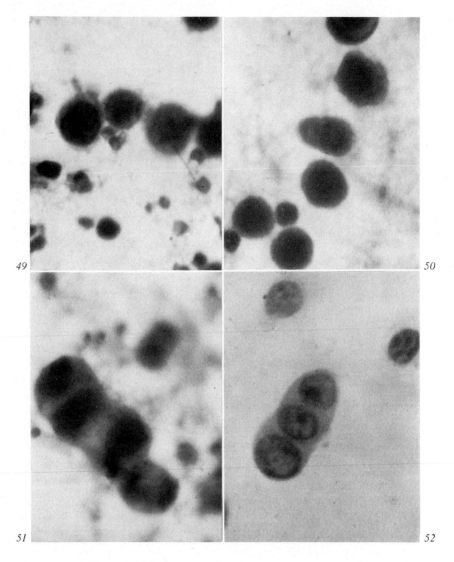

Fig. 49. Carcinoma of the breast. Pleural fluid. Small row of undifferentiated malignant cells suggestive of carcinoma of the breast. Papanicolaou. × 880.

Fig. 50. Carcinoma of the breast. Pleural fluid. Row of small undifferentiated malignant cells with little cytoplasm. Papanicolaou. × 880.

Fig. 51. Carcinoma of the breast. Pleural fluid. Four malignant cells in single file characteristic for scirrhous carcinoma. Papanicolaou. × 550.

Fig. 52. Carcinoma of the breast. Pleural fluid. Short row of nonvacuolized small malignant cells, indicative of scirrhous carcinoma of the breast. Papanicolaou. × 1,100.

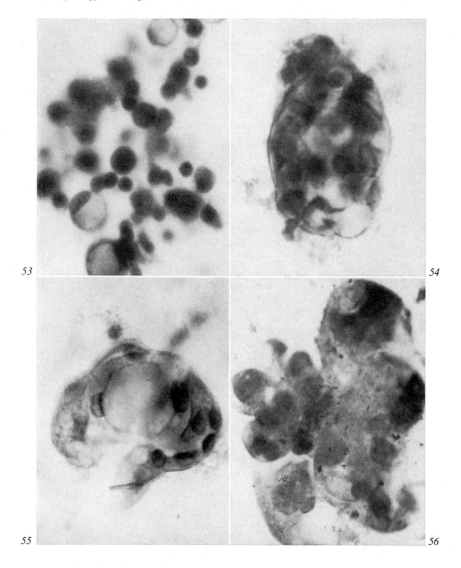

Fig. 53. Carcinoma of the breast. Pleural fluid. Clumps of small malignant cells with edematous mesothelial cells. Papanicolaou. × 550.

Fig. 54. Carcinoma of the breast. Pericardial fluid. Cluster of mucus-containing adenocarcinoma cells. Papanicolaou. × 550.

Fig. 55. Carcinoma of the breast. Pericardial fluid. Groups of adenocarcinoma cells with large amounts of mucus. Papanicolaou. × 550.

Fig. 56. Carcinoma of the breast. Pleural fluid. One group of large adenocarcinoma cells with foamy cytoplasm and malignant nuclei. Papanicolaou. × 440.

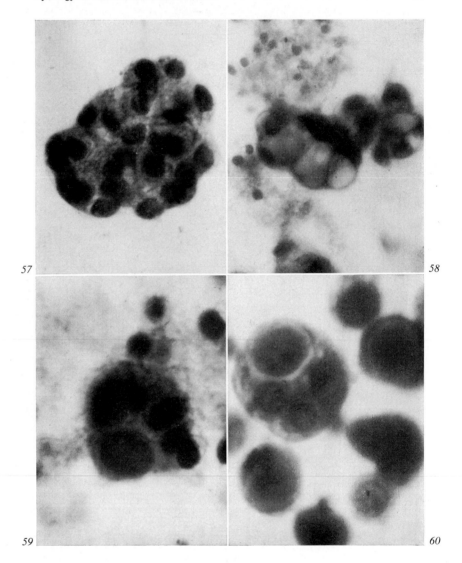

Fig. 57. Carcinoma of the breast. Pleural fluid. Clone of small undifferentiated malignant adenocarcinoma cells. Papanicolaou. × 550.

Fig. 58. Carcinoma of the breast. Pleural fluid. Groups of adenocarcinoma cells with large cytoplasmic vacuoles. Papanicolaou. × 550.

Fig. 59. Carcinoma of the breast. Pleural fluid. Small group of undifferentiated malignant cells. Papanicolaou. × 1,100.

Fig. 60. Carcinoma of the breast. Pleural fluid. Group of undifferentiated adenocarcinoma cells with inclusions. Papanicolaou. × 1,100.

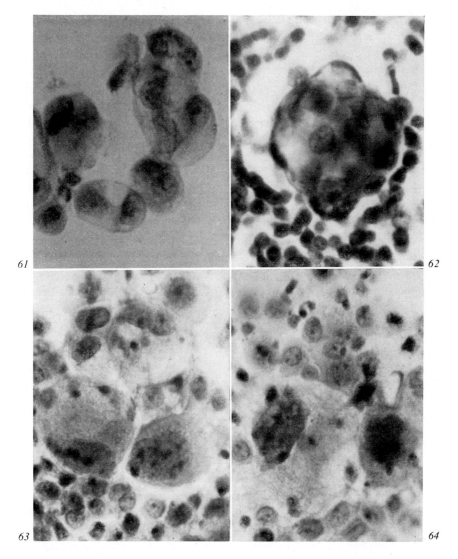

Fig. 61. Carcinoma of the breast. Pleural fluid. Large vacuolated adenocarcinoma cells. Papanicolaou. × 550.

Fig. 62. Carcinoma of the breast. Pleural fluid. Cluster of adenocarcinoma cells with a large amount of mucus. Papanicolaou. × 700.

Fig. 63. Carcinoma of the breast. Pleural fluid. Two large malignant cells showing the effect of chemotherapy. Papanicolaou. × 440.

Fig. 64. Carcinoma of the breast. Pleural fluid. Two large malignant cells showing the effect of chemotherapy. Papanicolaou. × 440.

ductular carcinoma of the breast. Figure 59 represents a cell group of un-differentiated or medullary carcinoma of the breast which is one of the most common metastasizing forms of breast carcinoma. The cells are so undiffer-entiated that only the history can give a clue as to the origin of those cells. Figure 60 represents a similar group of undifferentiated cells of medullary cancer of the breast. The cell group in the center is engulfing a large inclusion body, the origin of which is unknown. Similar inclusion bodies have been found by us in experimental cancers of the breast and were interpreted as viral inclusion bodies. Figure 61 represents cells of a large cell adenocarcinoma of the breast with many multinucleated cells which have distinct nucleoli and large mucous vacuoles in the cytoplasm. They are the cell type which we associate with the usual type of adenocarcinoma of the breast. Figure 62 shows a conglomeration of similar cells to form a clump of adenocarcinoma cells mixed with mucous which often obstructs the large breast ducts from which they can be extruded by pressure. This is an example of a so-called comedo cell carcinoma of the larger breast ducts. Figures 63 and 64 represent groups of adenocarcinoma cells of the breast showing the effect of chemo-therapy. The cells are enlarged and the nuclei in figure 64 are extremely pyknotic. There is a very severe histiocytic host reaction present.

Carcinoma of the Lung

It is only natural that carcinoma of the lung metastasizes frequently to the pleural cavities. Carcinoma of the lung is one of the most common malignancies in both sexes and although it has been linked with the habit of smoking, there are many variants of carcinoma of the lung which have no connection with this habit. Carcinoma of the lung arises from the bronchial tree including the mainstem bronchi down to the smaller bronchioles. For this reason, the varieties of metastatic carcinoma of the lung are so great.

Figure 65 shows a small clump of malignant cells representing a small un-differentiated carcinoma of the lung. Those cells originate probably from the basal cells or reserve cells of the bronchial epithelium, and this type of tumor is probably not connected with the smoking habit. Notice that the malignant cells which form a small cluster show pyknotic nuclei with large nucleoli and that they are still larger than the lymphocytes of this field. Figure 66 shows a similar picture with a centrally located small cluster of undifferentiated malignant cells varying in size and structure. A similar cell group of small

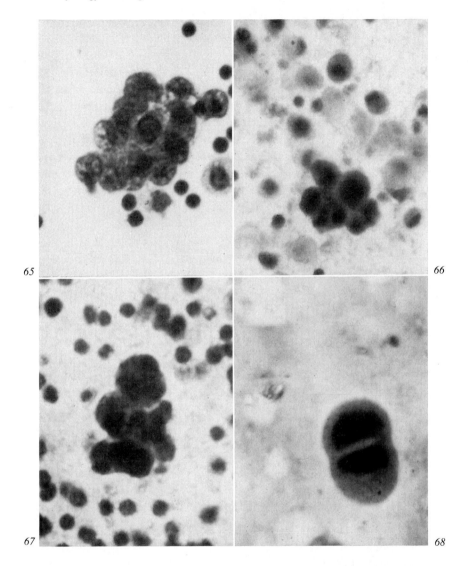

Fig. 65. Carcinoma of the lung. Pleural fluid. Group of small undifferentiated malignant cells. Papanicolaou. × 700.

Fig. 66. Carcinoma of the lung. Pleural fluid. Groups of small malignant cells suggestive of oat cell carcinoma. Papanicolaou. × 550.

Fig. 67. Carcinoma of the lung. Pleural fluid. Group of undifferentiated small malignant cells suggestive of oat cell carcinoma. Papanicolaou. × 880.

Fig. 68. Carcinoma of the lung. Pleural fluid. Small group of poorly differentiated malignant cells of the small cell type. Papanicolaou. × 1,100.

undifferentiated cells can be seen in figure 67. The cells in all three pictures show a great similarity and are representative of the small cell undifferentiated carcinoma of the lung which also carries the name of oat cell carcinoma. Figure 68 shows a larger magnification of two cells of similar type. Notice the compact structure of the nucleus without distinguishable nucleolus and the homogenous appearance of the cytoplasm which does not have any vacuoles. Figure 69 shows malignant undifferentiated cells from a carcinoma of the lung which are much larger than the oat cell carcinoma shown previously. They may represent malignant cells developing from an incompletely metaplastic squamous cell epithelium and we designate them as undifferentiated large cell carcinoma of the lung. A similar group of cells are found in the ascites fluid of figure 70. Here too, the cells are larger than in figures 65–68 and show a greater variety in the structure of the nuclei and in the appearance of the cytoplasm which shows numerous small vacuoles. The cell group in figure 71 shows a medium-sized cell size with little evidence of differentiation and also belongs to the group of undifferentiated carcinomas of the lungs. Figure 72 shows an isolated small group of medium-sized undifferentiated malignant cells accompanied by large numbers of reactive mesothelial cells. Figure 73 shows a coherent row of malignant cells with a distinct malignant nucleus and pink-staining cytoplasm. There is some indication of some intercellular bridges between individual cells. We feel this represents poorly differentiated squamous cell carcinoma of the lung in which the carcinoma developed after previous squamous cell metaplasia of the bronchial mucosa. Figure 74 shows a single malignant cell of the squamous cell type representative of the squamous cell carcinoma of the lung which appears well-differentiated and highly keratinized. The cell is very large and the cytoplasm shows numerous fibrils of keratin. Figure 75 shows a group of four cells, with large nuclei and with scanty cytoplasm with little evidence of keratinization. Nevertheless, the shape of the cells and their grouping characterize them as squamous cell carcinoma. One of the outstanding examples of a malignant squamous cell is depicted in figure 76. The cell is binucleated and is very large. The cytoplasm has distinct borders and is filled with kerato-hyalin.

While undifferentiated carcinomas of the small and large types and squamous cell carcinomas are the most frequent types of lung tumors leading to metastasis in the pleural cavities, the group of adenocarcinoma must not be forgotten. We differentiate two types of adenocarcinomas: the large cell type of adenocarcinoma originating from the glands located in the wall of the larger bronchi and the small cell adenocarcinomas which develop in

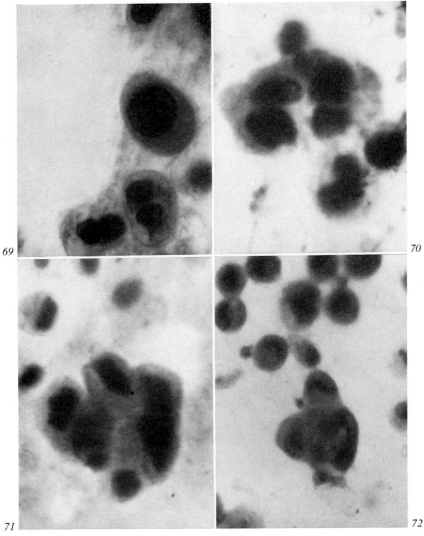

Fig. 69. Carcinoma of the lung. Pleural fluid. Clumps of poorly differentiated large malignant cells with few vacuoles. Papanicolaou. × 880.

Fig. 70. Carcinoma of the lung. Ascites. Group of medium-sized undifferentiated malignant cells. Papanicolaou. × 1,100.

Fig. 71. Carcinoma of the lung. Pleural fluid. Small group of poorly differentiated malignant cells of the medium-sized cell type. Papanicolaou. × 1,100.

Fig. 72. Carcinoma of the lung. Pleural fluid. Small clumps of undifferentiated malignant cells scattered among large numbers of reactive mesothelial cells. Papanicolaou. × 550.

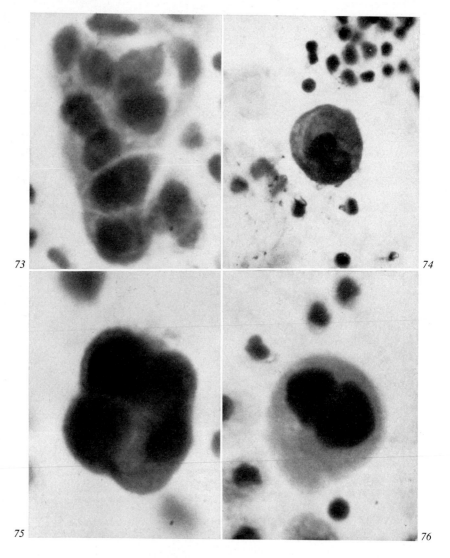

Fig. 73. Carcinoma of the lung. Pleural fluid. Group of malignant cells resembling squamous cell carcinoma. Papanicolaou. × 1,100.

Fig. 74. Carcinoma of the lung. Pleural fluid. Single highly keratinized malignant squamous cell. Papanicolaou. × 550.

Fig. 75. Carcinoma of the lung. Pleural fluid. Group of four malignant squamous cells with little keratinization. Papanicolaou. × 1,100.

Fig. 76. Carcinoma of the lung. Pleural fluid. Large malignant squamous cell with two nuclei. Papanicolaou. × 1,100.

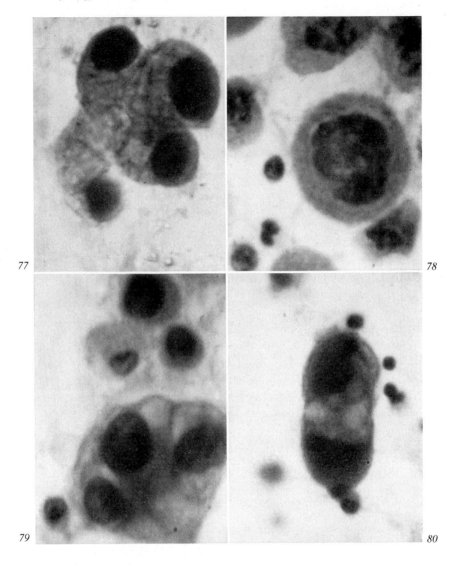

Fig. 77. Carcinoma of the lung. Pleural fluid. Group of poorly differentiated adeno-carcinoma cells. Papanicolaou. × 1,100.

Fig. 78. Carcinoma of the lung. Pleural fluid. Single undifferentiated adenocarcinoma cell. Papanicolaou. × 1,100.

Fig. 79. Carcinoma of the lung. Pleural fluid. Clumps of well-differentiated large adenocarcinoma cells with many vacuoles. Papanicolaou. × 550.

Fig. 80. Carcinoma of the lung. Pleural fluid. Two large malignant cells with many mucus vacuoles suggestive of adenocarcinoma. Papanicolaou. × 550.

the periphery and which are called alveolar carcinoma. Figure 77 is a group of medium-sized cells containing large pyknotic nuclei and a vacuolic cytoplasm. We designated them as adenocarcinoma of the lung arising from the glands of the larger bronchi. A similar cell is depicted in figure 78. The cell is giant-sized with a highly malignant nucleus and comparatively little mucous in the cytoplasm. Figure 79 shows clumps of well-differentiated cells imitating gland formation. The nuclei are large and dark-staining; the cytoplasm is filled with large vacuoles containing mucous. This gland has even a lumen in its center. Figure 80 shows two large malignant cells and the cytoplasm is filled with mucous vacuoles. They undoubtedly represent parts of a large cell adenocarcinoma.

Small cell adenocarcinomas of the lung are much less common and develop usually in the periphery of the lobes of the lung. They have no connection with the smoking habit of the patient and their peculiarity consists that the cancer cell invades and spreads throughout the lung tissue without destroying the alveolar septa. Figure 81 represents a group of small cells with malignant nuclei and a foamy cytoplasm. We designated them as small cell adenocarcinoma. Figure 82 contains a small group of cells with large pyknotic bean-shaped nuclei and vacuolized cytoplasm. It is this appearance of the cytoplasm which differentiates these cells from the cells of figures 66 and 67 and makes them small adenocarcinoma cells. Figure 83 gives an especially good example of this type of lung cancer. We see a compact group of small cells having a papillary shape and most of the cells containing mucous secretion. Figure 84 is a larger cluster of similar cells and the mucous content in each cell is even more prominent. The compactness of the group and the cell size characterizes them as alveolar cell carcinoma of the lung.

Carcinoma of the Gastrointestinal Tract

Tumors of the gastrointestinal tract which metastasize to the serous cavities are the carcinoma of the esophagus, the carcinoma of the stomach, and the carcinoma of the large intestines. The carcinomas of the esophagus are usually squamous cell carcinomas. They metastasize by direct migration of the malignant cells into the pleural and peritoneal cavities or by lymph vessel metastasis. Figure 85 shows a group of pale malignant cells with large nuclei and multiple nucleoli. They are gathered around a larger cell of similar characteristics. Those are cells from a squamous cell carcinoma of the lower third of the esophagus which have produced an ascites with nodular spread

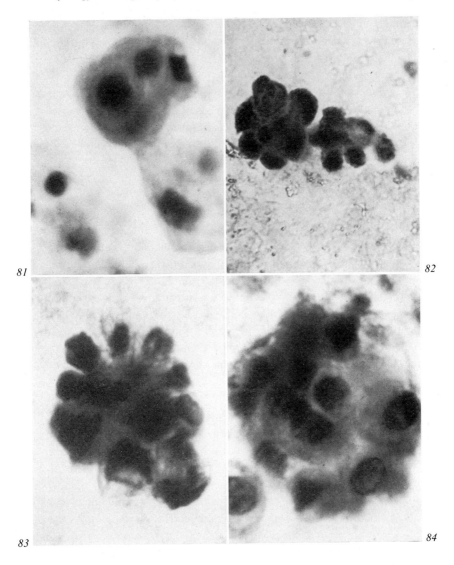

Fig. 81. Carcinoma of the lung. Pleural fluid. Group of small malignant cells characteristic of small cell adenocarcinoma. Papanicolaou. × 1,100.

Fig. 82. Carcinoma of the lung. Pleural fluid. Cluster of small adenocarcinoma cells
with bean-shaped nuclei and vacuolic cytoplasm. Papanicolaou. × 550.

Fig. 83. Carcinoma of the lung. Pleural fluid. Cluster of small adenocarcinoma cells
with evidence of secretion. Papanicolaou. × 700.

Fig. 84. Carcinoma of the lung. Pleural fluid. Cluster of small malignant adenocarcinoma cells with some evidence of secretion. Papanicolaou. × 1,100.

throughout the peritoneal cavity. Figure 86 represents a large malignant squamous cell which is multinucleated and contains a rich amount of keratin. This cell was taken from a pleural fluid of a patient with inoperable carcinoma of the middle of the esophagus.

Carcinomas of the stomach are all adenocarcinomas of different degrees of differentiation. Some cells are very small and contain little or no mucous while others form clusters of large adenocarcinoma cells. The peculiar thing about carcinoma of the stomach is that the cells usually appear singly and cluster formations are rare. A good example of this is figure 87, where we notice very loose groups of adenocarcinoma cells with little mucous, large nuclei and a granular cytoplasm without evidence of secretion of mucous (fig. 88). Figure 89 shows a small group of single malignant cells with prominent nucleoli. Figure 90 shows a single malignant cell surrounded by some small lymphocytes without evidence of any mucous secretion.

Figure 91 is a good example of the other type of carcinoma of the stomach, the mucous-secreting adenocarcinoma. We see a cluster of cells with large nuclei and multiple nucleoli surrounded by a cytoplasm which contains a large amount of mucous. A similar picture can be also seen in figure 92, which is taken from the pleural fluid. The cell group at the upper half of the picture contains a large amount of mucous which has pushed the nucleus toward the cell membrane, giving us the impression of a signet ring cell. The lower half of the picture shows three adenocarcinoma cells with a normal amount of mucous.

Carcinomas of the colon are 98% adenocarcinomas. They grow directly through the wall of the colon and then spread by means of the lymph vessels quickly through the peritoneal cavity. Most of the metastases occur in the peritoneal cavity, although occasionally they may metastasize in the other serous cavities. The fast-growing or undifferentiated types of adenocarcinomas produce clusters of small cells with highly pyknotic nuclei and a small rim of cytoplasm which does not give a mucous stain (fig. 93). As the tumors become more differentiated, the cells increase in size and the cytoplasm shows the accumulation of mucous. Instead of simple clusters, they form tubular or glandular forms imitating the glands of the colon. In figure 94, we see a mixture of poorly differentiated carcinoma cells and cells with varying amounts of mucous secretion. Sometimes, the production of mucous by the carcinoma is so extensive that the ascites assumes a mucoid-like appearance and large amounts of free mucous can be found inspissated in the omentum or the wall of the intestines. We then speak of colloid carcinoma of the colon. Such an example of an early colloid carcinoma can be seen in

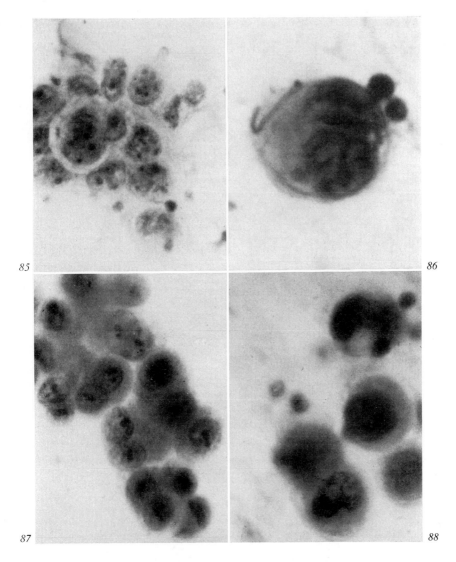

Fig. 85. Carcinoma of the esophagus. Ascites. Cluster of malignant cells with large nuclei and multiple nucleoli. Papanicolaou. × 550.

Fig. 86. Carcinoma of the esophagus. Pleural fluid. Large multinucleated malignant cell with keratinized cytoplasm. Papanicolaou. × 1,100.

Fig. 87. Carcinoma of the stomach. Pleural fluid. Loose groups of small adeno-carcinoma cells with dense cytoplasm and multiple nucleoli. Papanicolaou. × 550.

Fig. 88. Carcinoma of the stomach. Pleural fluid. Clumps of small undifferentiated malignant cells without secretion. Papanicolaou. × 1,100.

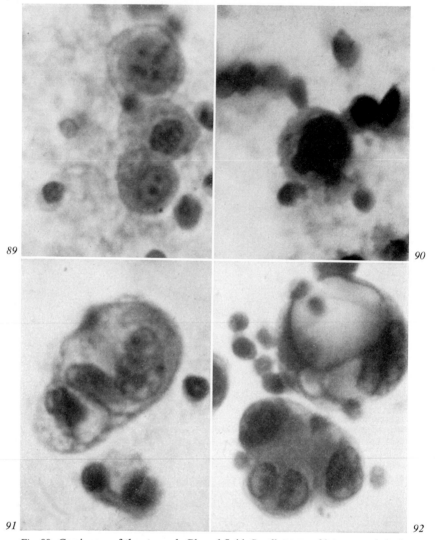

Fig. 89. Carcinoma of the stomach. Pleural fluid. Small group of large round single malignant cells with prominent and multiple nucleoli. Papanicolaou. × 1,100.

Fig. 90. Carcinoma of the stomach. Ascites. Single malignant cell without secretion suggestive of adenocarcinoma of the upper GI tract. Papanicolaou. × 1,100.

Fig. 91. Carcinoma of the stomach. Ascites. Large mucus-containing malignant cells with multiple nucleoli characteristic of adenocarcinoma of the GI tract. Papanicolaou. × 1,100.

Fig. 92. Carcinoma of the stomach. Pleural fluid. Groups of adenocarcinoma cells with a large amount of mucus. Papanicolaou. × 880.

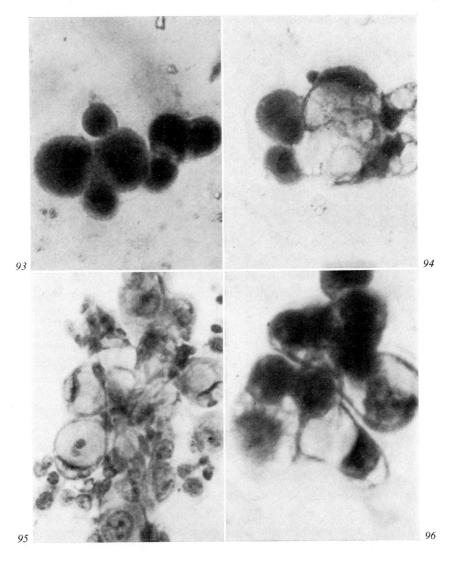

Fig. 93. Carcinoma of the colon. Ascites. Group of undifferentiated malignant adenocarcinoma cells without evidence of secretion. Papanicolaou. × 550.

Fig. 94. Carcinoma of the colon. Ascites. Five malignant cells attached to a large amount of mucus. Papanicolaou. × 550.

Fig. 95. Carcinoma of the colon. Ascites. Masses of cells with large vacuoles and signet ring-shaped nuclei. Papanicolaou. × 550.

Fig. 96. Carcinoma of the colon. Ascites. Groups of adenocarcinoma cells with large amounts of mucus. Papanicolaou. × 1,100.

figure 94, where a group of small poorly differentiated adenocarcinoma cells are attached to a large amount of mucous which probably comes from the vacuolated cells on the other side of the cell cluster. Figure 95 represents a papillary group of malignant cells from an ascitic fluid. This type of tumor is usually less malignant and has a tendency to form free-floating cell groups in the effusion. It is related to the papillary carcinoma of the large intestines and shows less invasive tendencies than the adenomatous carcinoma which infiltrates much faster and metastasizes earlier. The papillary form of carcinoma of the colon usually has its origin from a benign polyp of the colon and has a tendency to form tender multiple slim polyps which float in the lumen of the intestines. Only later do they infiltrate the wall of the intestines and become invasive. Figure 95 represents a papillary group of malignant colon cells aspirated from an ascitic fluid. The individual cells contain a large amount of mucous which gives them a signet ring appearance. The papillomatous masses hang loosely together and easily break up into smaller particles. Figure 96 demonstrates a group of medium-sized cells each of which is filled with large or small mucous vacuoles. These cells represent a well-differentiated adenocarcinoma of the colon and the mucous can be demonstrated by the mucicarmin stain. Figure 97 shows a well-formed gland of adenocarcinoma cells in the ascites produced by a metastasizing carcinoma of the lower rectum. It shows a clump of mucous-secreting cells together with cells resembling squamous cells. It is a well-differentiated adenocarcinoma and mucous can be demonstrated by the mucicarmin stain. Figure 98 represents a large group of cells with intercellular borders and no evidence of secretion. This represents a group of carcinoma cells taken from the ascites of a carcinoma of the colon demonstrating the phenomenon of cannibalism.

The lowest portion of the rectum is the anus. It contains squamous cells and mucous-secreting glandular cells and tumors arising in this portion are quite a common occurrence. Factors of chronic irritation such as chronic constipation and rectal fissures have been blamed for the frequency of carcinomas of the lower rectum. Microscopically, they can be either squamous cell carcinomas or adenocarcinomas. Sometimes, both cell types participate in the tumor formation and this tumor is then called adenoacanthoma. Figure 99 is a group of cells aspirated from the Douglas space between the rectum and the bladder. We see a cell clump which in one part of the picture resembles a mucinous adenocarcinoma cell, while in the other portion, the cells are closely packed and dense and suggest malignant squamous cells. Figure 100 represents a cell block preparation from an ascites of a well-differentiated carcinoma of the colon with well-formed glands in combina-

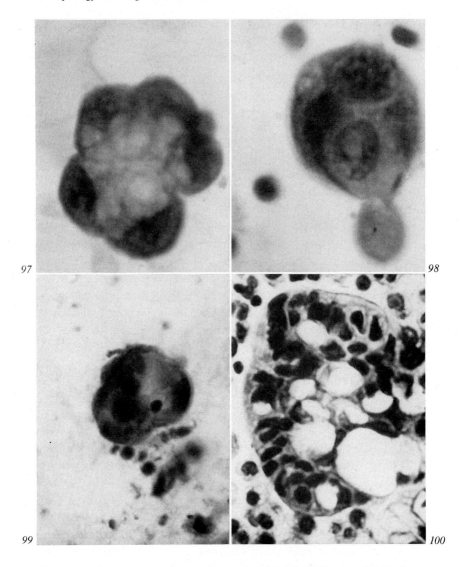

Fig. 97. Carcinoma of the colon. Ascites. A well-formed gland composed of adeno-carcinoma cells with large mucus vacuoles. Papanicolaou. × 1,100.

Fig. 98. Carcinoma of the colon. Ascites. Small clump of adenocarcinoma cells demonstrating cannibalism. Papanicolaou. × 1,100.

Fig. 99. Carcinoma of the rectum. Ascites. Clumps of mucus secreting malignant cells forming glands and papilli. Papanicolaou. × 550.

Fig. 100. Carcinoma of the colon. Ascites. Metastatic gland with pyknotic nuclei and mucus secretion. Cell block preparation. × 550.

tion with small papilli. The nuclei have all the slender shape of the colon carcinoma and the glands contain numerous vacuoles of mucous. This is a cell block preparation from an ascites and demonstrates best the spread of adenocarcinomas of the colon in the serous cavities.

Carcinoma of the Liver, Gallbladder, and Pancreas

Carcinomas of all three organs metastasize more frequently to lymph nodes or to other organs than to the serous cavities. If they do, they almost exclusively metastasize to the peritoneal cavity, causing an ascites with clumps of tumor cells. Figure 101 represents a hepatoma cell recovered from an ascites in a patient with liver cirrhosis. The cell has an outstanding size, is multinucleated, and each nucleus contains several nucleoli. The cytoplasm has a well-formed cell membrane and has a slightly foamy structure. The other cells in the same picture are mostly reactive mesothelial cells with few lymphocytes and histiocytes. Figure 102 represents a large hepatoma cell from an ascites in a patient who suffered from a solitary hepatoma. The cell is truly of giant size; it has numerous nuclei and multiple nucleoli. Adjoining this cell, a small hepatoma cell can be noted. The cytoplasm in the large hepatoma cell is quite compact and possesses only few small vacuoles. Some villi can be observed at the cell membrane.

Figure 103 represents a group of adenocarcinoma cells taken from an ascites of a patient with carcinoma of the gallbladder. The cells are adeno-carcinoma cells in good glandular formation; they are small and possess small dark nuclei with a vacuolated cytoplasm. Figure 104 is an unusual cell cluster of malignant adenocarcinoma cells taken from the pleural fluid of a patient with carcinoma of the gallbladder. The cells are from a cell block preparation and show a compact group of small adenocarcinoma cells. There is some mucous secretion in the cytoplasm.

Carcinomas of the pancreas are characterized by their late clinical symp-toms and their late evidence of metastasis. The patients are usually ailing for some time from a gastrointestinal disorder which hides the true nature of the disease for some time. Microscopically, they present all variations from undifferentiated small cell carcinomas arising from the acini of the gland to well-formed adenocarcinomas originating from the larger or smaller pan-creatic ducts. Carcinomas of the head of the pancreas are often associated with early painless jaundice, due to an obstruction of the common bile duct or the papilla of Vater, while carcinomas of the body or tail of the pancreas

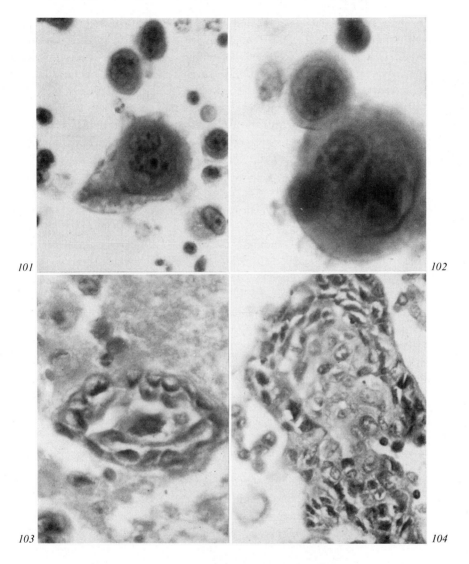

Fig. 101. Hepatoma. Ascites. Large malignant multinucleated cell with multiple nucleoli and a foamy cytoplasm. Papanicolaou. × 550.

Fig. 102. Hepatoma. Ascites. Giant-sized malignant multinucleated cell resembling a liver cell with multiple nucleoli. Papanicolaou. × 880.

Fig. 103. Carcinoma of the gallbladder. Ascites. Gland formed by middle-sized adenocarcinoma cells. Papanicolaou. × 550.

Fig. 104. Carcinoma of the gallbladder. Pleural fluid. Clump of malignant cells with some mucus secretion. Cell block preparation. × 265.

have no jaundice and remain symptomless for a long time. Malignant island cell tumors metastasize rarely, but they are outstanding by their hormone production. Figure 105 represents an undifferentiated small cell carcinoma of the pancreas with cells resembling a small cell carcinoma of the stomach. Note the marked inflammatory reaction in this ascites. Figure 106 comes from a carcinoma of the pancreas metastasizing to the pleural cavity. We see pale large malignant cells with bizarre nuclei and vacuolic cytoplasm. Numerous mesothelial cells are present. Figure 107 represents a well-differentiated adenocarcinoma of the pancreas with large cells arising from one of the larger ducts. Notice the large nuclei with very prominent nucleoli and the indistinct cytoplasm rich in vacuoles. Figure 108 represents a carcinoma of the pancreas which had metastasized to the pleura. We see two large malignant adenocarcinoma cells separated by a large amount of mucous. Figure 109 represents an undifferentiated small cell carcinoma of the pancreas similar to figure 105. However, the cells are larger, more pyknotic, and they possess a vacuolized cytoplasm. Figure 110 has been taken from the pleural fluid and represents a conglomerate of small and middle-sized malignant cells with large undifferentiated nuclei and a varying amount of cytoplasm. Those cells represent a poorly differentiated small cell adenocarcinoma. Figure 111 represents cells from an ascites showing clumps of undifferentiated malignant cells with large nuclei and ill-defined slightly foamy cytoplasm. Figure 112 has been taken from a pleural fluid and represents a giant-sized binucleated adenocarcinoma cell with distinctly foamy cytoplasm.

Islet cell tumors of the pancreas also may be malignant and metastasize to the peritoneal cavity. Beta cell tumors are credited with the production of insulin and induce clinically significant hypoglycemia due to hyperinsulinism. The alpha cells of the islets produce glucagon which induces severe degrees of hyperacidity with the formation of ulcers in the stomach and small intestines. It is probable that the islet cell disorders are connected with severe developmental disturbances. The malignant variants of those cells produce an occasional ascites where the specific cells cannot be identified.

Carcinoma of the Genitourinary Tract

Malignant tumors of the genitourinary tract, including the male sex organs, metastasize widely throughout the human body. They also cause effusions in all serous cavities and large quantities of malignant cells can be found in the exudates produced by the metastatic lesions. The malignant

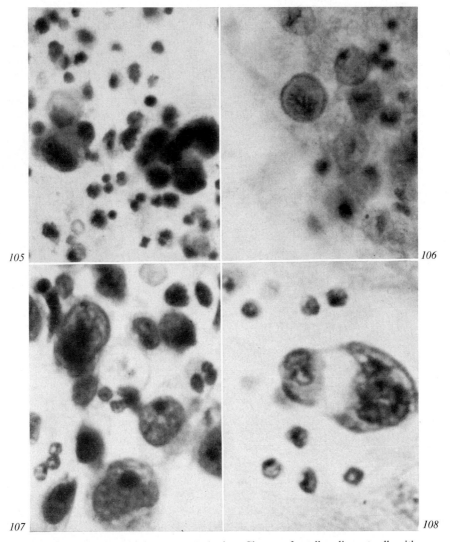

Fig. 105. Carcinoma of the pancreas. Ascites. Clumps of small malignant cells with prominent nucleolus and foamy cytoplasm. Papanicolaou. × 550.

Fig. 106. Carcinoma of the pancreas. Pleural fluid. Single large malignant cell with bizarre nucleus and foamy cytoplasm surrounded by mesothelial cells. Papanicolaou. × 550.

Fig. 107. Carcinoma of the pancreas. Ascites. Groups of malignant cells with small vacuoles and giant nucleoli. Papanicolaou. × 880.

Fig. 108. Carcinoma of the pancreas. Pleural fluid. Two malignant adenocarcinoma cells separated by large amounts of mucus. Cell block preparation. × 1,100.

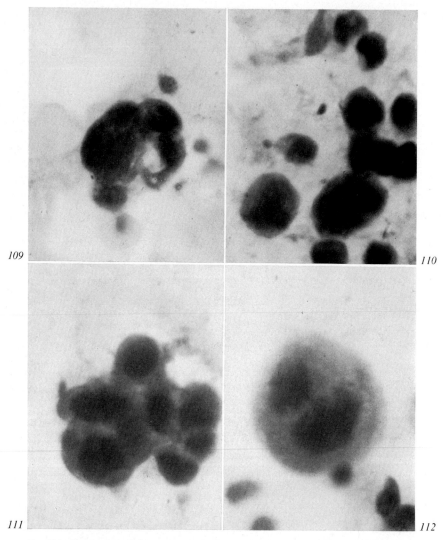

Fig. 109. Carcinoma of the pancreas. Ascites. Groups of poorly differentiated adeno-carcinoma cells with little secretion. Papanicolaou. × 550.

Fig. 110. Carcinoma of the pancreas. Pleural fluid. Small groups of undifferentiated malignant cells resembling undifferentiated small cell adenocarcinoma. Papanicolaou. × 1,100.

Fig. 111. Carcinoma of the pancreas. Ascites. Clumps of adenocarcinoma cells without evidence of secretion. Papanicolaou. × 1,100.

Fig. 112. Carcinoma of the pancreas. Pleural fluid. Gigantic binucleated adeno-carcinoma cell. Papanicolaou. × 1,100.

tumors of the kidney have been called renal cell carcinoma or hypernephroma. They start as small solid tumors in the kidney parenchyma, but soon penetrate into the blood vessels and metastasize widely throughout the body. In the blood vessels they grow as solid tumor masses and the favorite sites of metastasis are the lungs. They are composed of masses of anaplastic dark-staining cuboidal cells and so-called clear cell tumors which form glandular masses.

Figure 113 is a group of cells taken from an ascites of a patient with tubular cell carcinoma of the left kidney. The cells possess irregular dark-staining nuclei with multiple nucleoli and a homogenous cytoplasm. Figure 114 is a small group of malignant tubular cells taken from the pleural fluid which show the features between the anaplastic and the clear cell type of hypernephroma cells. Figure 115 shows a group of cells taken from the pleural fluid which are the typical clear cell type of carcinoma. Note the pale cytoplasm, the relatively small pale nuclei, and the cluster formation. In figure 116, we have one of each cell type. The larger one is characterized by its foamy cytoplasm and its many villi on the cell membrane. The smaller one resembles a cell in figure 113. The cells are small, with a large dark nucleus and a homogenous staining cytoplasm.

Other malignant tumors from the kidneys arise in the pelvis as transitional cell carcinomas. They infiltrate the local fat tissue but do not metastasize extensively or produce serous exudates. A tumor of embryonal origin develops occasionally in the kidneys of very young children and carries the name of embryonal carcinoma of the kidney or Wilms tumor. These tumors do metastasize widely, but seldom produce malignant effusions.

Cancers of the urinary bladder do seldom metastasize to the serous cavities. We differentiated between the squamous cell carcinoma, the much less common adenocarcinoma, and the much more common transitional cell carcinoma. The metastases of carcinoma of the bladder occur usually by direct penetration of the carcinoma cells through the bladder wall into the peritoneal cavity. Figure 117 shows single malignant transitional cells in the ascites fluid which are undifferentiated and have a vacuolated cytoplasm.

Carcinomas of the prostate also do not metastasize frequently in the serous cavities, but prefer to infiltrate the surrounding tissue of the pelvic structures. Figure 118 shows nests of small cells with pyknotic and irregular nuclei and a bare rim of cytoplasm. This tumor had metastasized to the pleural cavity. It resembles in many respects the oat cell carcinoma of the lung. Figures 119–124 show that carcinomas of the prostate do possess the capacity of differentiation and form true adenocarcinomas with the cells

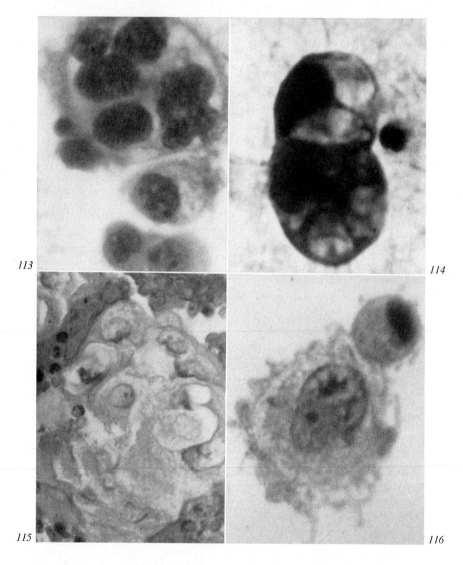

Fig. 113. Hypernephroma of the kidney. Ascites. Large dark cells with multiple nucleoli. Papanicolaou. × 880.

Fig. 114. Carcinoma of the kidney. Pleural fluid. Malignant pyknotic cells, some of which are highly vacuolated. Papanicolaou. × 1,100.

Fig. 115. Carcinoma of the kidney. Pleural fluid. Large vacuolated malignant clear cells. Papanicolaou. × 550.

Fig. 116. Hypernephroma of the kidney. Ascites. Two malignant clear cells with foamy cytoplasm and multiple nucleoli. Papanicolaou. × 880.

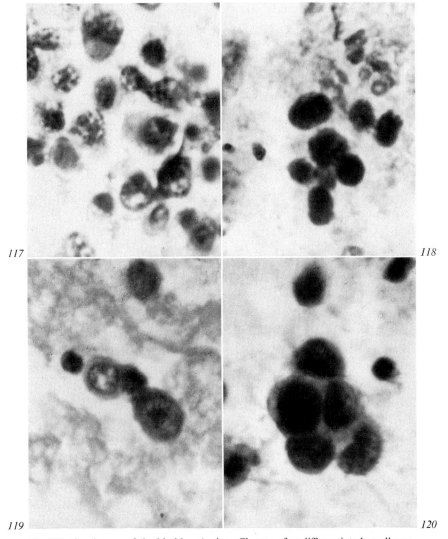

Fig. 117. Carcinoma of the bladder. Ascites. Clumps of undifferentiated small non-secreting malignant cells. Cell block preparation. × 880.

Fig. 118. Carcinoma of the prostate. Pleural fluid. Nests of small cells with pyknotic nucleus and very little cytoplasm suggestive of small cell carcinoma. Papanicolaou. × 1,100.

Fig. 119. Carcinoma of the prostate. Pleural fluid. Group of small adenocarcinoma cells. Cell block preparation. × 1,100.

Fig. 120. Carcinoma of the prostate. Pleural fluid. Cluster of small undifferentiated adenocarcinoma cells. Papanicolaou. × 1,100.

arranged like glands and with varying amounts of cytoplasm. They also contain PAS-positive material in the lumen. Figure 120 is probably the least differentiated adenocarcinoma of the group, although it is more differentiated than figure 118, in which not even the cell type of an adenocarcinoma can be identified. Figure 121 shows a fairly well-organized clump of malignant cells with intercellular bridges. Figure 122 is representative of the so-called small cell adenocarcinoma of the prostate which has a tremendous invasive power and metastasizes to the brain, the bone, and all serous cavities. Figure 123 is also a well-differentiated adenocarcinoma from the pleural fluid showing a rather well-formed gland made up of malignant cells with a dark-staining mass in the center of the lumen. Figure 124 is part of an adenocarcinoma of the prostate with rows of well-differentiated adenocarcinoma cells.

Tumors of the testis which metastasize to the serous cavities are the seminoma and embryonal carcinoma of the testis. The origin of the seminoma is not quite known, but it is the most common tumor of the testis and is composed of groups of pale round cells with prominent and multiple nucleoli. The tumor cells are infiltrated with numerous small lymphocytic cells. The significance of those small lymphocyte-like cells is not known. Some authors regard them as abortive forms of spermatazoa. Figure 125 shows a cell cluster from an ascites caused by a seminoma which had metastasized to the abdominal lymph nodes. Notice the regular-sized cells, the prominent nucleoli and the relatively pale-staining cytoplasm. A similar group of cells showing similar features is depicted in figure 126. Figure 127 shows a cell group of much less well-differentiated seminoma cells. Lymphocyte-like structures which are attached to or engulfed by some of the cells are also evident. Figure 128 represents a picture from a metastasizing embryonal carcinoma of the testis. Those tumors are highly malignant and invade very quickly, joining structures. The cells are quite irregular in shape and the lymphocytic structures are absent. Whenever embryonal carcinoma of the testis metastasizes, one can never be certain which of the malignant components of the tumor will metastasize. Most frequently, the ectodermal parts of the tumor will metastasize and we can find metastatic carcinomas, hepatomas, or brain tumors. Sometimes, however, the mesodermal tissues will metastasize and we then find spindle cell sarcomas or rhabdomyosarcomas or chondrosarcomas. Occasionally, two different tissues will participate on the metastatic process and we will find tumors with mixed type of metastasis.

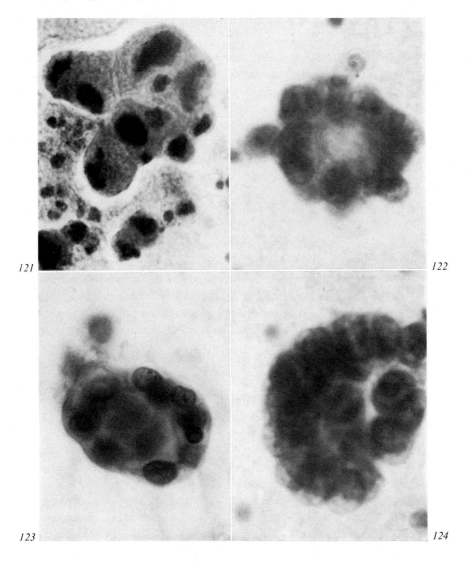

Fig. 121. Carcinoma of the prostate. Pleural fluid. Glands formed by malignant cells characteristic of small cell adenocarcinoma. Papanicolaou. × 550.

Fig. 122. Carcinoma of the prostate. Pleural fluid. Small gland formed by malignant cells with little cytoplasm. Papanicolaou. × 880.

Fig. 123. Carcinoma of the prostate. Pleural fluid. Single gland formed by small adenocarcinoma cells. Papanicolaou. × 550.

Fig. 124. Carcinoma of the prostate. Pleural fluid. Glandular arrangement of small undifferentiated malignant cells. Papanicolaou. × 1,100.

Carcinoma of the Cervix and Endometrium

Carcinoma of the cervix is usually a squamous cell carcinoma, although in rare instances an adenocarcinoma may develop from the cervical glands. Carcinoma of the cervix is probably the best-studied carcinoma in the human body, and extensive experiments have shown that its development is always preceded by a lesion called carcinoma *in situ*. The difference between carcinoma *in situ* and the invasive type of carcinoma of the cervix is that carcinoma *in situ* does not metastasize, while the invasive carcinoma metastasizes early using the lymph vessels of the uterus and the pelvis as the principal pathways of spreading. Carcinoma of the cervix will produce peritoneal metastasis causing an ascites containing malignant cells, but it will also metastasize to the other serous cavities such as the pericardial cavity. In the pelvis, it forms firm masses of carcinoma which encroach upon the vital structures of the pelvis, the ureters and sigmoid, leading to intestinal obstruction and uremia. We differentiated various degrees of maturity in carcinoma of the cervix: completely undifferentiated forms which like other similar undifferentiated forms possess no distinctive characteristics for this type of tumor, and well-differentiated forms which are all squamous cell carcinomas with evidence of severe keratinization and the formation of pearly bodies. Figure 129 represents a group of small keratinized squamous cells with distinct borders and large pyknotic nuclei. They are surrounded by histiocytes. The cells have been taken from a pleural fluid with metastasizing carcinoma of the cervix. Figure 130 represents a cell group of a squamous cell carcinoma of the cervix metastasizing to the peritoneal cavity. Notice the large malignant cells with bizarre shapes and abundant amounts of highly keratinized cytoplasm. Figure 131 represents a group of poorly differentiated adenocarcinoma cells of the cervix which has metastasized to the pleural cavity. The cells are round and small and the cytoplasm possesses a small amount of mucous. Figure 132 represents a cell group from an adenocarcinoma of the cervix in the pleural fluid. The cells are much larger and possess large amounts of pale-staining mucous.

Carcinomas of the endometrium are usually adenocarcinomas. Sometimes, malignant squamous cells will be participating in the neoplastic process and then we call the lesion adenoacanthoma. In a rare instance, carcinomas of the endometrium will contain a mixture of malignant cartilage cells, fibrosarcoma cells, and malignant muscle cells. This then represents a mixed mesodermal tumor of the endometrium. Figure 133 shows four undifferentiated small cells with large nuclei and no evidence of mucous secretion.

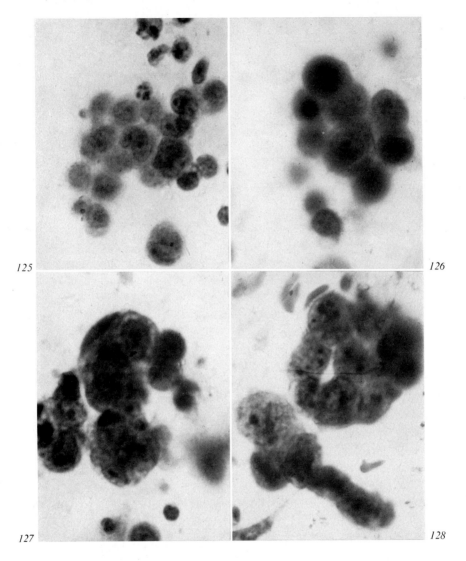

Fig. 125. Seminoma. Ascites. Groups of round cells with prominent nucleoli together with small cells resembling lymphocytes. Papanicolaou. × 550.

Fig. 126. Seminoma. Ascites. Large single cells with prominent nucleoli. Papanicolaou. × 550.

Fig. 127. Seminoma. Ascites. Group of large malignant glandular cells mixed with small cells resembling lymphocytes. Papanicolaou. × 880.

Fig. 128. Carcinoma of the testes. Pleural fluid. Gland formed by small malignant cells. Papanicolaou. × 700.

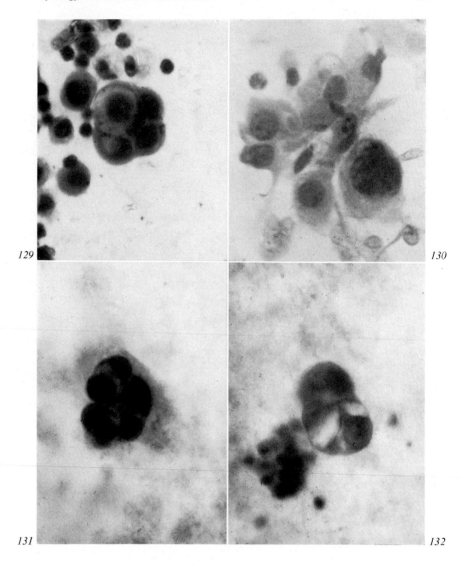

Fig. 129. Carcinoma of the cervix. Pleural fluid. Cluster of small malignant keratinized squamous cells. Papanicolaou. × 550.

Fig. 130. Carcinoma of the cervix. Ascites. Groups of well-keratinized malignant squamous cells. Papanicolaou. × 550.

Fig. 131. Carcinoma of the cervix. Pleural fluid. Gland formed by malignant cells with large nucleoli and foamy cytoplasm. Papanicolaou. × 550.

Fig. 132. Carcinoma of the cervix. Pleural fluid. Group of adenocarcinoma cells with vacuolated cytoplasm and prominent nucleoli. Papanicolaou. × 550.

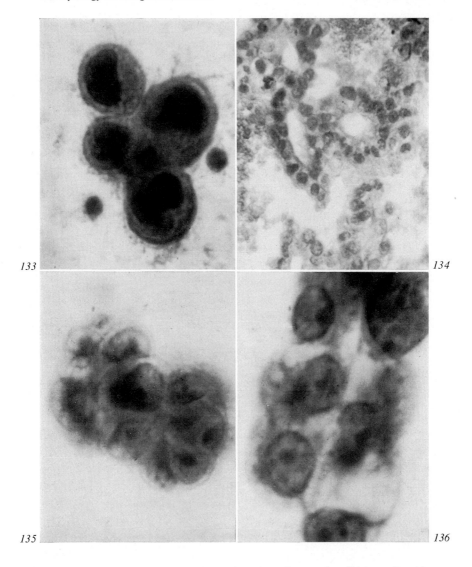

Fig. 133. Carcinoma of the endometrium. Ascites. Group of malignant cells with little evidence of secretion. Papanicolaou. × 880.

Fig. 134. Carcinoma of the endometrium. Ascites. Well-formed glands lined by small malignant cells. Cell block preparation. × 880.

Fig. 135. Carcinoma of the endometrium. Pleural fluid. Cluster of pale foamy cells with large nucleoli. Papanicolaou. × 880.

Fig. 136. Carcinoma of the endometrium. Pleural fluid. Glands formed by large malignant adenocarcinoma cells. Papanicolaou. × 1,100.

These are typical of the undifferentiated carcinoma of the endometrium. Figure 134 is a much smaller magnification of a glandular picture of adenocarcinoma of the endometrium with well-formed glands lined by small malignant cells. Figure 135 has been taken from a pleural fluid and represents a large cell adenocarcinoma with big pale nuclei, common nucleoli and a foamy cytoplasm. Figure 136 represents a similar tumor which metastasized to the pleura with large adenocarcinoma cells with multiple nucleoli and large amounts of mucous.

Carcinoma of the Ovaries

Ovarian carcinoma is one of the most frequent types of carcinoma metastasizing in the abdominal cavity where it produces large amounts of recurrent ascites. With proper treatment, patients can live with the disease an astounding number of years, although a complete cure of carcinoma of the ovary is not known. Malignant tumors of the ovary are divided into solid carcinomas and cystic carcinomas.

Solid carcinomas of the ovary are much less frequent than the cystic forms and can appear in the form of a solid adenocarcinoma, a solid papillary carcinoma or a solid medullary carcinoma. Other special varieties are the malignant granulosa cell carcinomas, arrhenoblastomas, and dysgerminomas. The latter three types are characterized by a special hormone production. Figure 137 represents a group of malignant granulosa cells metastasizing to the pleural cavity in an 86-year-old woman who showed definite estrogenic activity in her vaginal smear. The tumor cells are round and of the same size. They possess a pyknotic nucleus and a deep purple cytoplasm. They form small solid clusters but also appear as single cells in the pleural fluid. Another form of carcinoma of the ovary is the solid primary ovarian cancer which may occur in several histological varieties. Figure 138 represents the medullary type of ovarian carcinoma which shows groups of cells divided into lobules. The cells are large and do not show any evidence of secretion. Figure 139 presents an example of an alveolar carcinoma of the ovary where small groups of malignant cells are arranged in alveolar groups separated by connective tissue. Figure 140 represents the plexiform carcinoma of the ovary where the malignant cells are arranged in cords and papillary structures. They are undifferentiated which prevents them from being mistaken for malignant granulosa cells.

Cystic carcinomas of the ovaries are much more frequent and lead to the formation of very large cystic tumors which can originate in one or both

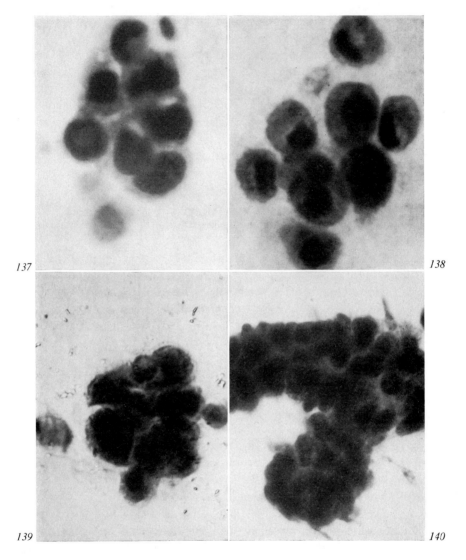

Fig. 137. Carcinoma of the ovary. Pleural fluid. Malignant granulosa cells with small pyknotic nucleus and little dense cytoplasm. Papanicolaou. × 1,100.

Fig. 138. Carcinoma of the ovary. Ascites. Clump of adenocarcinoma cells with little vacuolization. Papanicolaou. × 1,100.

Fig. 139. Carcinoma of the ovary. Ascites. Cluster of malignant cells with large nucleus and a small amount of cytoplasm. Papanicolaou. × 550.

Fig. 140. Carcinoma of the ovary. Ascites. Papillary structure formed by small un-differentiated malignant cells. Papanicolaou. × 550.

ovaries. The cystic tumors are usually multilocular and the individual cysts are separated by thick or thin septa. The cystic content is either a thin serous fluid or appears as a thick pseudomucinous material. Some of the smaller cysts contain blood. If the content is serous, we speak of serous cystadeno-carcinoma; if the content is mucinous, we speak of pseudomucinous cys-tadenocarcinoma. The tumor cells in the serous cystadenocarcinoma are small and cuboidal and form papillary masses which often fill the smaller cysts completely (fig. 141). They have the tendency to invade the capsule of the tumor and form small papillary tumors on the outside of the capsule. Histologically, they appear malignant, forming disorganized papillae which metastasize to the other ovary or the organs of the peritoneal cavity, where they form papillary implants. The effusions produced by the serous cys-tadenocarcinoma usually are quite massive and have a tendency to recur quickly after paracentesis.

Figure 142 represents a malignant gland from a case of serous cystadeno-carcinoma with multiple nuclei and evidence of cannibalism. In figure 143, which comes from a pleural fluid, the malignant cells are larger and the foamy cytoplasm is characteristic of a malignant serous cystadenocarcinoma. Figure 144 depicts a group of malignant glands with bizarre shapes and a cytoplasm filled with foamy material which is probably serous fluid.

The other type of cystic malignant tumor is the pseudomucinous cystade-nocarcinoma. These also form large cystic tumors with a smooth outer surface and many multilocular cysts filled with slightly greenish and often coagulated material. The cyst walls are lined with tall epithelium resembling cervical epithelium. The malignant degeneration of this type of tumors is much less frequent than in the serous cystadenocarcinoma. The cells assume the character of a well-differentiated adenocarcinoma with many cell layers and an increased amount of mitosis. Invasion of the capsule of the tumor by atypical and malignant glandular structures is frequent.

Figure 145 represents a small cluster from a case of pseudomucinous cystadenocarcinoma with many undifferentiated cells and thick mucous droplets in the other cells. Figures 146–148 represent cell samples from metastasizing pseudomucinous cystadenocarcinomas recovered from pleural fluid or ascites. Figure 146 shows very large cells with pyknotic nuclei and a large amount of mucous in the form of big vacuoles. Figure 147 shows smaller cells with malignant nuclei pushed towards the margins of the cells and large amounts of mucous droplets in the cytoplasm. Figure 148 repre-sents a pleural metastasis of a pseudomucinous cystadenocarcinoma with many cells filled with large amounts of mucous.

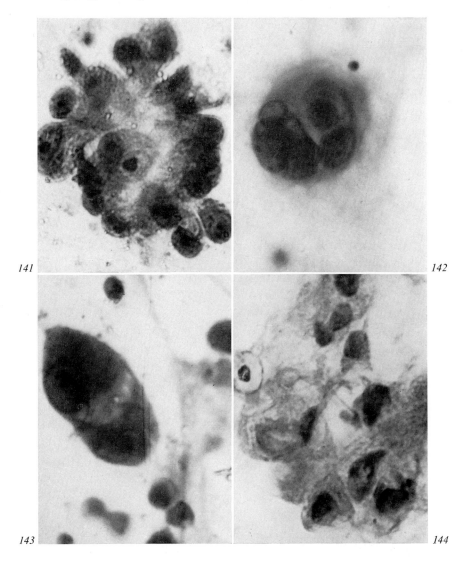

Fig. 141. Carcinoma of the ovary. Ascites. Gland of malignant cells with a large amount of cytoplasm and prominent nucleoli. Papanicolaou. × 550.

Fig. 142. Carcinoma of the ovary. Ascites. Clump of malignant cells with little mucus and cannibalism. Papanicolaou. × 550.

Fig. 143. Carcinoma of the ovary. Pleural fluid. Two large malignant cells with large nuclei and compact cytoplasm. Papanicolaou. × 1,100.

Fig. 144. Carcinoma of the ovary. Ascites. Group of large malignant cells with abundant foamy cytoplasm and bizarre nuclei. Papanicolaou. × 550.

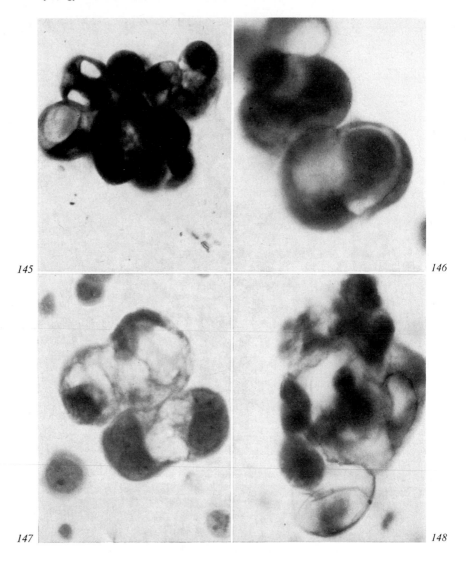

Fig. 145. Carcinoma of the ovary. Pleural fluid. Cluster of malignant cells containing large nuclei and mucus vacuoles. Papanicolaou. × 700.

Fig. 146. Carcinoma of the ovary. Ascites. Clumps of malignant cells with large nuclei and large amounts of mucus. Papanicolaou. × 1,100.

Fig. 147. Carcinoma of the ovary. Ascites. Group of malignant cells with large vacuoles of mucus. Papanicolaou. × 700.

Fig. 148. Carcinoma of the ovary. Pleural fluid. Malignant cells with large amounts of mucus in the cytoplasm. Papanicolaou. × 550.

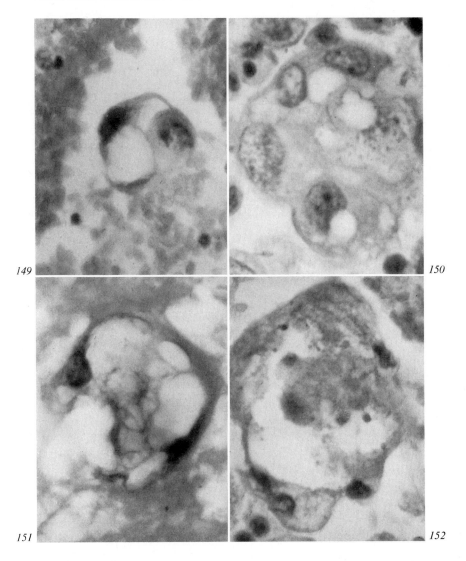

Fig. 149. Carcinoma of the ovary. Pleural fluid. Malignant cells with large mucus vacuoles and spindle-shaped nucleus. Cell block preparation. × 550.

Fig. 150. Carcinoma of the ovary. Ascites. Large malignant cells with vacuolated cytoplasm in gland formation. Cell block preparation. × 880.

Fig. 151. Carcinoma of the ovary. Ascites. Malignant cells showing signet ring shape in the cytoplasm distended by mucus. Papanicolaou. × 700.

Fig. 152. Carcinoma of the ovary. Ascites. Large degenerated adenocarcinoma cell filled with mucus-like material. Papanicolaou. × 700.

Degenerative forms of cystadenocarcinoma cells are frequent. They are characterized by giant cell size, signet ring cells and a cytoplasm which is completely filled with mucous vacuoles. Figure 149 shows one of the cells completely filled with mucous. Figure 150 shows a cluster of cells, each one containing a large amount of mucous. Figure 151 shows a degenerated group of cells in which the mucous apparently has exuded from the cytoplasm and penetrated the surrounding tissues. Figure 152 shows a similar cell from an ascites with mucous exuding from the degenerated cytoplasmic membrane.

Metastasizing Sarcomas

The 120 cases that we have discussed in this chapter represented all metastasizing epithelial tumors which produced exudates in the serous cavities. Malignant mesodermal tumors also metastasize in the serous cavities and produce effusion containing malignant cells. However, metastases of sarcomas are much less common and we have encountered only a few examples of such an occurrence. Figure 153 represents cells from an ascites of a 14-year-old boy who developed a massive tumor in the liver region. The tumor originateb in the liver capsule and spread throughout the peritoneal cavity causing considerable ascites. The cells of figure 153 are spindle cells with large nuclei and distinct nucleoli. The cells vary considerably in size and some multinucleated large cells can also be recognized. Figure 154 is taken from the same tumor and shows one multinucleated large cell with an abundant amount of cytoplasm containing many intracytoplasmic fibrils. This is embryonal rhabdomyosarcoma of the peritoneal cavity with fibrillary cytoplasm and large cells known as strap cells. The boy succumbed to this disease.

Figure 155 represents a cluster of large cells with malignant nuclei and many vacuoles. It was taken from an ascites fluid from a tumor which developed in the mesentery. Figure 156 is a large single cell from the same tumor with a marginal nucleus and cytoplasm completely replaced by a large vacuole. Using Sudan-3, the vacuoles were proved to be neutral fat which designated the cell as a liposarcoma cell. Liposarcomas are unusual tumors which originate in fat tissue. Various degrees of differentiation have been described ranging from spindle cell tumors with fine fat droplets in the cytoplasm to mature liposarcomas with large cells containing fat vacuoles. The tumor is slow-growing and with proper therapy can be cured.

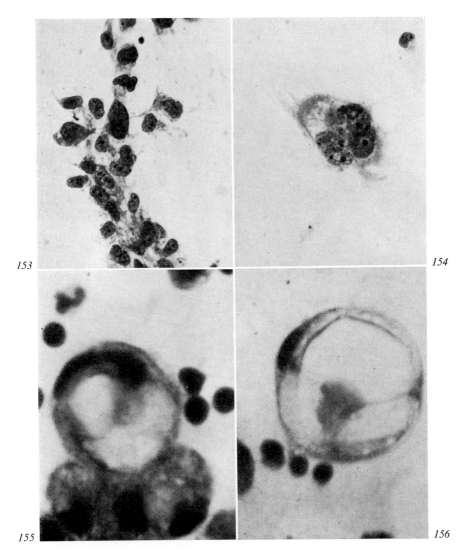

Fig. 153. Embryonal rhabdomyosarcoma. Ascites. Loosely textured stellate tumor cells with large nucleus and distinct nucleolus. Papanicolaou. × 550.

Fig. 154. Embryonal rhabdomyosarcoma. Ascites. Strap cell containing two to four nuclei and longitudinal intracytoplasmic fibriles. Papanicolaou. × 550.

Fig. 155. Liposarcoma. Ascites. Group of large malignant cells containing fat vacuoles. Papanicolaou. × 880.

Fig. 156. Liposarcoma. Ascites. Single large malignant cell filled with fat. Papanicolaou. × 1,100.

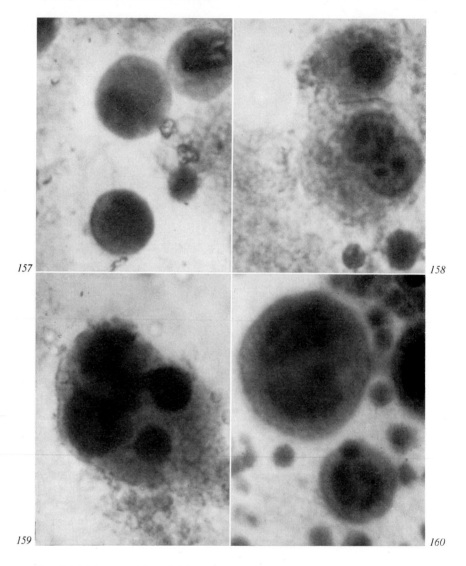

Fig. 157. Melanoma. Pleural fluid. Three melanoma cells with large nuclei and melanin pigment. Papanicolaou. × 1,100.

Fig. 158. Melanoma. Pleural fluid. Melanoma cells without pigment with many nucleoli. Papanicolaou. × 880.

Fig. 159. Melanoma. Pleural fluid. Cluster of melanoma cells with dark nucleus and no pigment. Papanicolaou. × 1,100.

Fig. 160. Melanoma. Ascites. Large round malignant cells with large nuclei and little pigment. Papanicolaou. × 1,100.

Melanoma

Figures 157–160 are devoted to a very malignant tumor, the melanoma. These are tumors which originate usually in the skin, but sometimes also in other mucous membranes or the eyes. They form pigmented tumors which metastasize quickly and cause extensive exudates in the serous cavities. The tumor cells are either round or spindle-shaped and show a varying amount of brown pigment which can be identified chemically as melanin. Sometimes, the melanin content is so heavy that all other structures of the cells are obscured while at other times very few or no melanin granules can be observed and the diagnosis of the tumor rests upon its cellular configuration. Figure 157 shows three melanoma cells with the pigment near the nuclei. Figure 158 shows two large melanoma cells, one of which is binucleated without any pigment. Figure 159 shows a cluster of melanoma cells from the pleural fluid with multiple nucleoli and a sparse amount of melanin. Figure 160 shows large melanoma cells from the ascites fluid with dark pyknotic nuclei and little melanin in the cytoplasm.

VII. The Cytology of Exudates in Malignant Blood Dyscrasias

Pleural or peritoneal exudates are a frequent occurrence in patients with malignant blood dyscrasias. In fact, according to MELAMED [140], the appearance of malignant blood cells in effusions may represent the initial manifestation of the disease. He examined 200 effusions in lymphoma patients and found malignant cells in about 45% of his cases. Most of the cells were found in pleural fluids. In our material, we found lymphosarcomas the most prevalent malignant blood disease to be followed by reticulum cell sarcoma and Hodgkin's disease. We were usually unable to differentiate between the cells of lymphosarcomas and the acute or chronic leukemias. Among the lymphosarcomas, we recognize the large blast cell type and the small lymphocytic type. In some patients, these lymphocytes show a marked degree of maturity while in others, the lymphocytes vary greatly in size with atypical nuclei and large or multiple nucleoli. Figure 161 represents large and immature lymphocytes from the ascites of a lymphosarcoma of the stomach. Figure 162 represents pleural fluid from a lymphosarcoma arising in the mediastinum. Here the cells are quite irregular in size with many bean-shaped or indented nuclei and a varying amount of cytoplasm. Figure 163 represents cells which MELAMED [140] called the small blast type. They have a uniform cell pattern and a dense cellularity. The cells often resemble benign lymphocytes as seen commonly in tuberculosis. Figure 164 shows much larger cells which originated from a lymphosarcoma of the thymus. The cells vary greatly in size and are larger than mature lymphocytes. The nuclei have a dense chromatin content and often multiple nuclei. Most cells have a well-defined large round nucleus with atypical nucleoli which may be single or multiple. They resemble immature lymphocytes and have been given the name large blast type lymphosarcoma cells by MELAMED [140]. They have a pale-staining cytoplasm which often shows small extrusions and usually appear together with a mixed background of other lymphocytic cells.

Quite different from the cells of lymphosarcoma or leukemia are the cells which arise from the reticulum cell sarcoma which also belongs to the family of malignant lymphoma cells. They are the cells which form the

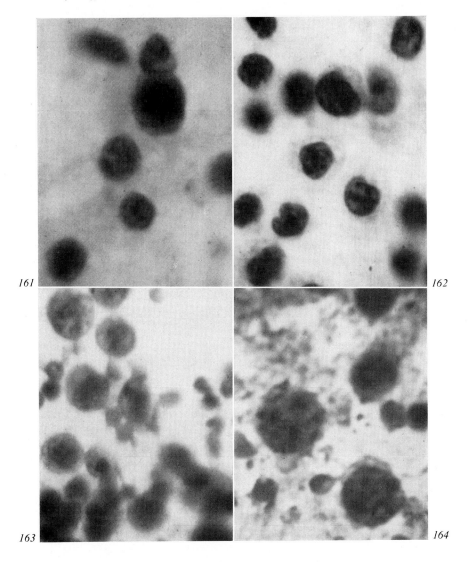

Fig. 161. Lymphoma. Ascites. Cells of the lymphoid group of various sizes and shapes consistent with lymphosarcoma. Papanicolaou. × 1,100.

Fig. 162. Lymphosarcoma. Pleural fluid. Malignant lymphocytes with prominent and multiple nucleoli and a varying amount of cytoplasm. Papanicolaou. × 1,100.

Fig. 163. Lymphosarcoma. Pleural fluid. Malignant lymphoma cells of the lymphoblastic type. Papanicolaou. × 1,100.

Fig. 164. Lymphosarcoma. Pleural fluid. Large malignant lymphoma cells with multiple nucleoli. Papanicolaou. × 1,100.

matrix of the lymphoid structures and are characterized by the ability to form a fine interconnecting network of reticulum which can be demonstrated with silver stains. They are called the reticulum cells and the malignant tumors of the reticulum cells form the reticulum cell sarcomas. The cells are usually larger than lymphocytes and have a highly irregular nuclear structure with many chromatin centers and large nucleoli. They all possess a palestaining cytoplasm which may be smooth or granular (fig. 165). Figure 166 demonstrates the great variability in the appearance of the nuclear structures which often have small nuclear extrusions. Some of the cells seem to be connected by a fine cytoplasmic reticulum which can be demonstrated by special silver stains. Figure 167 shows single malignant reticulum cells with demonstrable nuclear extrusions. The nuclei of those cells show a very coarse chromatin with prominent chromocenters and multiple highly atypical nucleoli. The cytoplasm of those cells varies greatly in amount, and some of the cells form a fine interconnecting reticulum. Figure 168 has been taken from the pleural fluid of a patient who received large doses of Cytoxan, the chemotherapeutic agent for malignant lymphomas. Prolonged use of this drug has a profound effect on the epithelium of various tissues leading to changes resembling a cancerous state. Koss [112] feels that those alkylating agents not only affect the RNA and DNA contents of the cells but other cell components as well. Notice the shrunken pale appearance of the nuclei with fading chromatin content and the washed-out appearance of the cells due to cytoplasmic degeneration.

Hodgkin's disease is a form of malignant lymphoma which frequently invades the serous cavities causing severe serous effusions. The disease usually originates in the lymph nodes or the spleen which form the source of the malignant cells. Sometimes, Hodgkin's disease also metastasizes to the serous membranes like any other malignant tumor. The cells of effusions caused by Hodgkin's disease are atypical lymphocytes, eosiniphile leukocytes and peculiar large multinucleated endothelial cells which have been first described by STERNBERG and which are now called Reed-Sternberg cells. The course of the disease follows many patterns. It may spread throughout the body, forming progressive granulomas in all organs, or it may remain restricted to one region of the body such as the lymph nodes of the neck. In most cases, the patient with Hodgkin's disease dies of the febrile granuloma. Some patients develop a truly invasive reticulum cell sarcoma which metastasizes and spreads like any malignant lymphoma. Involvement of the bone marrow usually causes a progressive anemia followed death. Figure 169 has been taken from an ascites of a patient whose mesentery lymph nodes

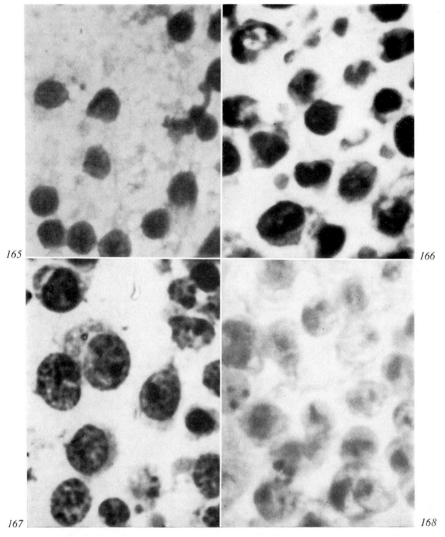

Fig. 165. Reticulum cell sarcoma. Pleural fluid. Small malignant cells with irregular shape resembling reticulocytes. Cell block preparation. × 880.

Fig. 166. Reticulum cell sarcoma. Pleural fluid. Malignant reticulum cells together with reactive mesothelial cells. Cell block preparation. × 1,100.

Fig. 167. Reticulum cell sarcoma. Ascites. Single small malignant cells with numerous nucleoli and a small rim of cytoplasm. Papanicolaou. × 1,100.

Fig. 168. Reticulum cell sarcoma. Pleural fluid. Groups of single malignant cells with cytoplasmic degeneration consistent with cytoxin therapy. Cell block preparation. × 1,100.

and spleen were affected by the disease. Notice the many atypical lymphocytes, many with two nuclei, and the large multinucleated endothelial cells
suggestive of Reed-Sternberg cells. Figure 170 shows a higher magnification
of the Reed-Sternberg cell in a cell block preparation. It shows the multinucleation and the many atypical nucleoli. Figure 171 shows the same type
of cell taken from an ascites in a patient with abdominal Hodgkin's disease.
The multinucleation, with the many chromatin clumps and the multiple
nucleoli can be shown distinctly together with a small amount of cytoplasm.
Figure 172 shows the effect of chemotherapy (Busulfan) on the mesothelial
cells of the pleural fluid. This drug is primarily useful against leukemias.
It very often causes prolonged remissions, but it affects also the mesothelial
cells of other organs such as the lungs. The drug causes nuclear enlargement
with hyperchromasia. Its effect resembles somewhat that of chronic radiation dysplasia. Koss [112] feels that those changes are similar to spontane-
ously occurring pre-cancerous changes and he cites two patients who developed carcinoma of the vulva and one carcinoma of the cervix after 5
years of Busulfan therapy.

The thymus gland is an organ which is composed of lymphoid follicles,
reticulum cells, and special cell structures resembling clusters of epithelial
cells. The function of the organ is still ill-understood, but we know that it
supplies special types of lymphocytes which are immunologically helpful in
the combat of disease in children and probably acts as an endocrine gland
during our period of growth. The most common type of malignant tumors
arising in the thymus gland are lymphocytic lymphomas which form large
tumors in the mediastinum and which also cause pleural effusions. Figure
173 shows a cluster of small mature lymphocytic cells taken from the pleural
fluid of a child with a thymoma. Notice that the lymphocytes are uniform
and that they possess a highly pyknotic nucleus with very little cytoplasm.
Figure 174 is a cell cluster from a thymoma taken from the pleural fluid
after radiation therapy. Notice that the cells are enlarged and their cytoplasm contains many vacuoles which is common after radiation. Figure 175
represents a specimen from an ascites from a young woman who suffered
from agnogenic myeloid metaplasia. Her ascites fluid contains numerous
myeloid cells in different stages of maturation. The large cell in the middle
of the picture represents a mature megakaryocyte. Figure 176 represents a
picture in a similar case with fewer and less mature myelocytes and a well-
preserved megakaryocytoblast.

Multiple myeloma is a disease which primarily affects the bone marrow,
but it can produce metastasis in other organs and also produce exudates in

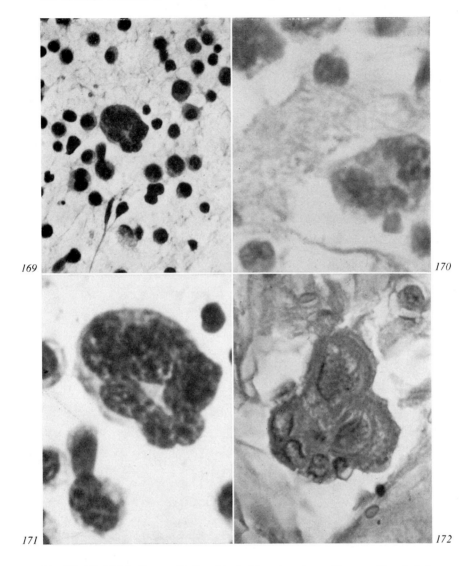

169

170

171

172

Fig. 169. Hodgkin's disease. Ascites. Atypical lymphocytes and large multinucleated cells suggestive of Reed-Sternberg cells. Papanicolaou. × 550.

Fig. 170. Hodgkin's disease. Ascites. Multinucleated Reed-Sternberg cell. Cell block preparation. ×1,100.

Fig. 171. Hodgkin's disease. Ascites. Large multinucleated cell characteristic of Reed-Sternberg cell surrounded by atypical lymphocytes. Papanicolaou. × 1,100.

Fig. 172. Hodgkin's disease. Pleural fluid. Group of bizarre mesothelial cells showing the effect of chemotherapy. Papanicolaou. × 700.

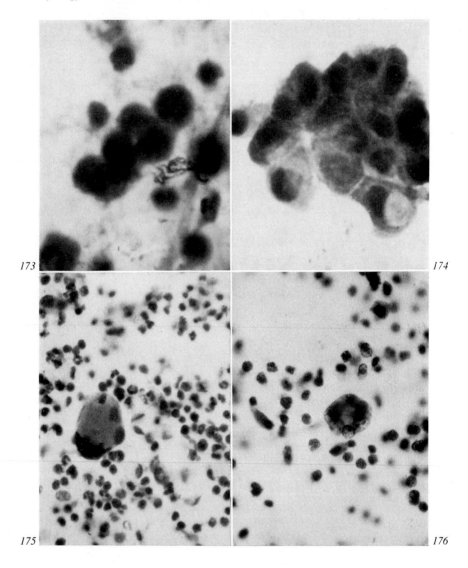

Fig. 173. Thymoma. Pleural fluid. Bloody background with many highly atypical lymphocytic-type cells. Papanicolaou. × 1,100.

Fig. 174. Thymoma. Pleural fluid. Clumps of small malignant cells resembling lymphocytes showing acute radiation changes. Papanicolaou. × 700.

Fig. 175. Myeloid metaplasia. Ascites. Large degenerated megakaryocytes with myeloid cells. Papanicolaou. × 550.

Fig. 176. Myeloid metaplasia. Ascites. Large megakaryocytes with numerous myeloid cells. Papanicolaou. × 550.

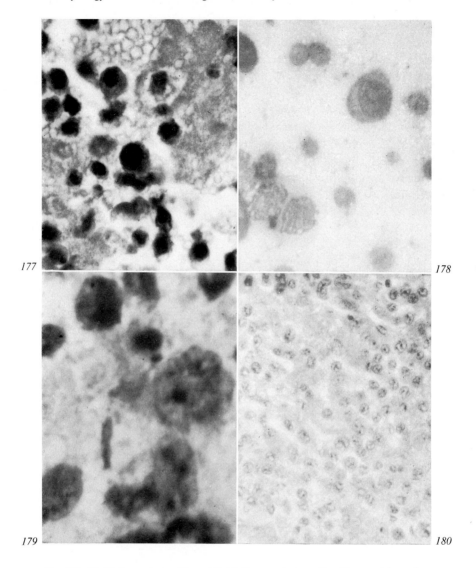

Fig. 177. Multiple myeloma. Pleural fluid. Few myeloma cells with eccentric nucleus and basophilic cytoplasm. Papanicolaou. × 550.

Fig. 178. Myeloma. Pleural fluid. Typical cells for plasma cell myeloma. Papanicolaou. × 880.

Fig. 179. Myeloma. Pleural fluid. Immature myeloma cells and one large myeloblast. Papanicolaou. × 1,100.

Fig. 180. Myeloma. Pleural biopsy. Infiltration with tumor of pleura composed of plasma cells. HE. × 265.

the serous cavities. The most common type of myeloma is the plasmocytic type in which the cells look like plasma cells with an eccentric nucleus and very distinct cytoplasm. Multiple myelomas produce an atypical protein which in turn causes severe kidney damage and death from uremia. Figure 177 shows a pleural fluid with few large myeloma cells characterized by the eccentric nucleus and the deep purplish cytoplasm. Figure 178 shows similar cells with high magnification. Figure 179 shows a pleural fluid with immature myeloma cells and several large myeloblasts with irregular nuclei and multiple nucleoli. Figure 180 shows a section from a metastatic lesion of a multiple myeloma in the pleura. Notice the homogenous appearance of the plasma cell-like myeloma cells.

VIII. The Effect of Chemotherapy and Radiotherapy upon the Cells of Transudates and Exudates

This chapter describes the effects of chemotherapy and radiation therapy in benign and malignant cells. The benign cells are represented mostly by mesothelial cells and occasionally by histiocytes. The malignant cells are represented by the metastatic tumor cells found in the exudates. Figures 181–188 represent chemotherapeutic changes in benign cells. Figures 189–192 represent radiation changes in benign mesothelial cells. Figures 193–196 represent chemotherapeutic changes in malignant cells, and figures 197–200 represent radiation changes in malignant cells.

In many instances, the cellular atypism produced by chemotherapy or radiation is so strong that it greatly reduces the morphological differences between benign mesothelial cells and metastatic malignant cells. The cells become enlarged, become vacuolized, and assume bizarre forms. The principal difference between radiation changes and chemotherapy changes are often of such fine degree that no sharp lines can be drawn between those differences, except maybe that the nuclei suffer more evidence of cell injury during radiation therapy than during chemotherapy, which primarily affects the cytoplasm [87].

Because of our experience with ascites cell tumors in mice, it seems only logical to utilize the chemotherapeutic and radiation effect on mesothelial cells as well as on malignant cells as an index of response to therapy. According to Koss [112], such a study of the response of various tumors to radiation and chemotherapy has not yet been performed, although some information on this subject is available. Malignant lymphoma cells for example respond to radiation or alkylation with nuclear fragmentation and karyorrhexis. Tumors growing in clusters of the papillary or glandular type respond usually very poorly to chemotherapeutic agents. One should follow the patient for at least 3 years before declaring him as cured. Changes due to radiation therapy and chemotherapy are remarkably alike although, as stated before, nuclei of the cells seem to be more affected by radiation therapy. Amongst the radionimetic drugs, the alkylating agents must be singled out since their effect on cancer cells resemble closest the radiation effect.

In addition to the appearance of atypical mesothelial cells, the disappearance of the malignant cells must be taken cautiously as a prognostically favorable sign [81]. Very often, this disappearance is only quite temporary and is followed by new showers of malignant cells. The reduction of metastatic tumor cells from effusions after radiation or chemotherapy is most marked in lymphomas and least marked in cases of metastatic adenocarcinomas. It is often accompanied by a decrease in the size of the effusion. However, in most cases those results are only temporary and do not indicate a permanent therapeutic effect.

Changes suggestive of chemotherapy effects are presented in figures 181–188. Figure 181 shows one large mesothelial cell with vacuolic cytoplasm and pyknotic nucleus with dense chromatin. In contrast to the normal mesothelial cells in the same picture, it represents a strong chemotherapy effect. Figure 182 represents the mesothelial cells from a pleural effusion in a case of gastric carcinoma which had metastasized to the pleura and the peritoneum. The cells are enlarged, the cytoplasm is vacuolized and the nuclei offer a picture of karyorrhexis with numerous fragments of nuclei within a vacuolized cytoplasm. Figure 183 represents mesothelial cells from a pleural fluid of a case of Hodgkins' disease which has been treated by chemotherapy for 3 years. Notice that the mesothelial cells are enlarged, the nuclei are bizarre and dense, and the cytoplasm shows numerous fine vacuoles. Figure 184 shows a single mesothelial cell in a case of advanced metastatic carcinoma of the lung showing a bizarre nucleus with clumping of chromatin and shrunken nuclear membrane. The cytoplasm contains numerous fine vacuoles. Figure 185 shows highly atypical mesothelial cells from a case of chronic myelogenous leukemia treated by chemotherapy. The leukemic cells can still be recognized in the picture. The mesothelial cells show marked nuclear enlargement with a small rim of dark-staining cytoplasm and greatly distorted cell shapes. Figure 186 shows mesothelial cells in the pleural fluid from a case of cancer of the breast which was treated for 2 years by chemotherapy. Notice atypical mesothelial cells with bizarre nuclei and a great variety of shapes. Figure 187 shows a cluster of cells from the pleural fluid in a case of metastatic carcinoma of the breast. The cells possess pyknotic nuclei, with foamy cytoplasm. Figure 188 shows mesothelial cells in the pleural fluid from a case of metastatic carcinoma of the breast. The cells are enormously enlarged and possess large nuclei which practically fill the cells completely. No nucleoli are present. Two cells in the same picture are normal mesothelial cells. Figure 189 shows cells from the pleural fluid of a case of Hodgkin's disease treated with radiation. The pic-

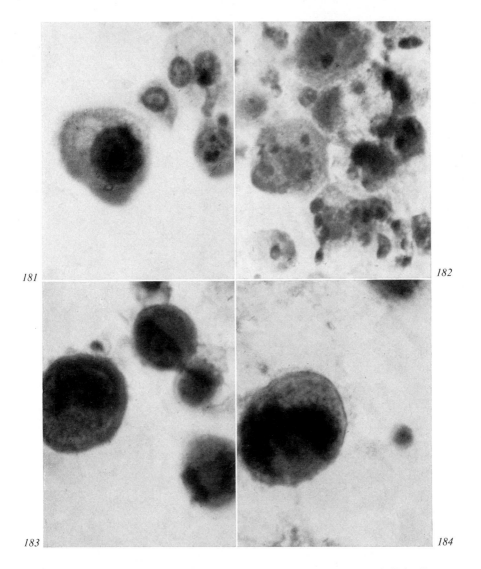

Fig. 181. Pleural fluid after chemotherapy for breast carcinoma. Mesothelial cells, histiocytes, large cells suggestive of chemotherapy. No malignant cells. Papanicolaou. × 880.

Fig. 182. Pleural fluid from a case of gastric carcinoma. Strong chemotherapy effect. Papanicolaou. × 880.

Fig. 183. Pleural fluid from a case of Hodgkin's disease showing strong chemotherapy effect involving mesothelial cells and plasma cells. Papanicolaou. × 880.

Fig. 184. Pleural fluid from a case of metastatic carcinoma of the lung, showing many single highly bizarre cells of mesothelial origin related to chemotherapy effect. Papanicolaou. × 880.

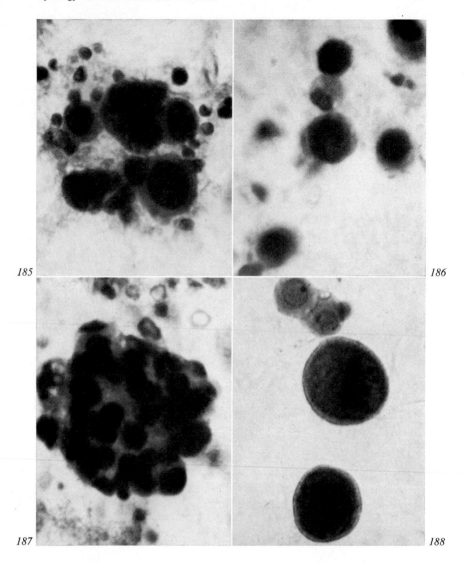

Fig. 185. Pleural fluid from a case of chronic myelogenous leukemia. Highly atypical mesothelial cells showing chemotherapy effect. Papanicolaou. × 880.

Fig. 186. Pleural fluid from a case after mastectomy and 2 years chemotherapy. Bizarre mesothelial cells showing strong chemotherapy effect. Papanicolaou. × 880.

Fig. 187. Pleural fluid from a case of metastatic carcinoma of the breast. Cluster of small malignant cells showing pronounced chemotherapy effect. Papanicolaou. × 880.

Fig. 188. Pleural fluid from a case of metastatic breast carcinoma. Large cells with indistinct chromatin indicative of chemotherapy effect. Papanicolaou. × 880.

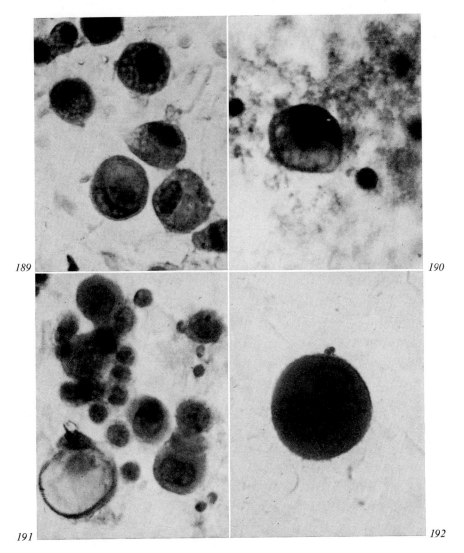

Fig. 189. Pleural fluid from a case of Hodgkin's disease treated with radiation. Group of atypical reticulum cells showing acute radiation effect. Papanicolaou. × 880.

Fig. 190. Pleural fluid from a case of metastatic carcinoma of the cervix. Large cell with vacuolized cytoplasm suggestive of acute radiation effect. Papanicolaou. × 880.

Fig. 191. Pleural fluid from a case of metastatic carcinoma of the pancreas. Many vacuolized cells with nuclei suggestive of acute radiation effect. Papanicolaou. × 880.

Fig. 192. Pleural fluid from a case of metastatic carcinoma of the breast. Single large cell with multiple nuclei suggestive of radiation effect. Papanicolaou. × 700.

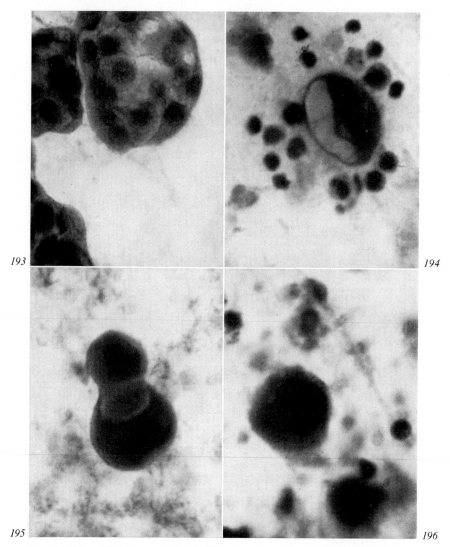

Fig. 193. Pleural fluid from a patient with metastatic carcinoma of the breast. Clump of cells with vacuolized cytoplasm consistent with chemotherapy effect. Papanicolaou. × 700.

Fig. 194. Pleural fluid from a case of metastatic reticulum cell sarcoma showing large malignant cells associated with chemotherapy. Papanicolaou. × 880.

Fig. 195. Pleural fluid from a case of metastatic carcinoma of the breast showing vacuolization and radiation cell changes. Papanicolaou. × 350.

Fig. 196. Pleural fluid from a case of metastatic carcinoma of the breast, showing malignant cells with radiation and chemotherapy effect. Papanicolaou. × 880.

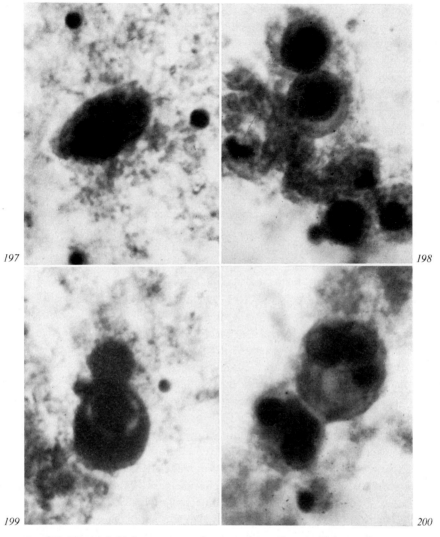

Fig. 197. Pleural fluid from a case of metastatic carcinoma of the cervix. Large atypical squamous cells with changes suggestive of radiation effect. Papanicolaou. × 555.

Fig. 198. Pleural fluid from a case of unknown primary carcinoma, showing numerous malignant cells with acute radiation effect. Papanicolaou. × 880.

Fig. 199. Pleural fluid from a case of metastatic carcinoma of the breast showing acute radiation effect.

Fig. 200. Pleural fluid from a case of squamous cell carcinoma of the lung. Large malignant cells with numerous vacuoles indicative of acute radiation effect. Papanicolaou. × 555.

ture shows atypical reticulum cells with small and shrunken nuclei and a highly vacuolized cytoplasm. This represents an acute radiation effect on benign cells. Figure 190 shows a large mesothelial cell from the pleural fluid of a case of squamous cell carcinoma of the cervix. The cell possesses a bizarre pyknotic nucleus and a large number of vacuoles. Figure 191 represents cells from the pleural fluid of a case of metastatic carcinoma of the pancreas. The cells show multinucleation and giant-sized vacuoles representing acute radiation effect. Figure 192 represents a mesothelial cell from the pleural fluid of a case of metastatic carcinoma of the breast. The cell is much enlarged and possesses three nuclei with crude chromatin structures. The plasma is deeply basophilic and represents an example of radiation effect on benign mesothelial cells.

Figure 193 shows a clump of cells from a patient with metastatic carcinoma of the breast. The pleural fluid contains numerous clumps of cells with relatively small nuclei but highly vacuolized cytoplasm. This represents the effect of chemotherapy on malignant cells. Figure 194 represents a large reticulum cell from metastatic reticulum cell sarcoma in the pleural fluid, showing vacuolization of the cytoplasm and bizarre enlargement of the cells. This represents predominantly a chemotherapy effect on a malignant reticulum cell. Figure 195 represents cells from a pleural fluid in metastatic carcinoma of the breast. It shows deeply pyknotic nuclei and large vacuoles between the cells showing a combination of chemotherapy and radiation changes. Figure 196 represents two malignant cells from the pleural fluid of a metastatic carcinoma of the breast. The nuclei are deeply pyknotic and fill completely the entire cell body. The cytoplasm is very scant, which represents a combination of chemotherapy radiation effect. Figure 197 is a large malignant squamous cell from the pleural fluid of a case of metastatic carcinoma of the cervix. The pyknotic nucleus with its rough chromatin structure is suggestive of radiation effect. Figure 198 shows malignant cells found in a pleural effusion in a case of malignancy of undetermined origin. The cells all show deeply pyknotic nuclei and highly vacuolized cytoplasm. This is suggestive of acute radiation effect on cells, probably belonging to the adenocarcinoma group. Figure 199 shows a large cell in the pleural fluid from a metastatic carcinoma of the breast. The nucleus is large and dense and the cell shows many vacuoles. Figure 200 shows two cells from a pleural fluid in a case of squamous cell carcinoma of the lung. The cells are malignant as shown by their large and bizarrely shaped nuclei. The cytoplasm contains numerous vacuoles suggestive of acute radiation effect.

IX. Analysis of the Author's Materials of Transudates and Exudates of the Serous Cavities

This chapter is devoted to an analysis of the material that we have used as the basis of our monograph. It was collected during our extensive comparative study on the accuracy of cancer cell detection by cytological methods during the years from 1965 to 1971. During this period, we examined 30,000 clinical and hospital patients for the detection of cancer cells. We were able to demonstrate that exfoliated carcinoma cells can be frequently detected in cases of carcinoma of the respiratory tract, the oral cavity, the stomach, and the urinary bladder with equal accuracy as in carcinoma of the cervix. Amongst our material, we analyzed also a large number of serous fluids which we examined for the presence or absence of malignant cells. Table I records the types of transudates and exudates in 1,052 patients who were examined. 140 specimens represented pericardial effusions, 563 specimens represented pleural effusions, and 349 represented peritoneal effusions.

The specimens were grouped into transudates, inflammatory exudates, and neoplastic exudates. Amongst the pericardial effusions, transudates re-

Table I. Types of transudates and exudates (total number of patients: 1,052)

	Number	Percentage
Pericardial effusions	140	100
Transudates	56	40
Inflammatory exudates	51	36
Neoplastic exudates	33	25
Pleural effusions	563	100
Transudates	181	32.1
Inflammatory exudates	142	25.2
Neoplastic exudates	240	42.6
Peritoneal effusions	349	100
Transudates	106	30.4
Inflammatory exudates	86	24.6
Neoplastic exudates	157	45

Table II. Distribution of sex and age amongst patients with effusions

	Number	Percentage
Sex		
Female patients	656	62
Male patients	396	38
Age (by group intervals of 10 years)		
0–10	12	1.1
11–20	27	2.4
21–30	31	3.1
31–40	50	4.9
41–50	151	14.8
51–60	306	29.0
61–70	334	31.7
71–80	118	11.2
over 81	23	2.2
Total	1,052	100

presented the largest number (40%), while neoplastic exudates represented the smallest number (25%). Amongst the pleural effusions, neoplastic exudates represented the largest number (42.6%), while inflammatory exudates represented the smallest number (25.2%). Amongst the peritoneal effusions, neoplastic exudates also represented the largest number (45%), while inflammatory exudates represented the smallest number.

Table II analyzes the distribution of sex and age among the patients with effusions in their serous cavities. The number of female patients was nearly twice as high as the number of male patients (62 vs. 38%), and the age of the patients varied a great deal according to our groups of 10-year intervals. We encountered only a very small number (1.1%) between the ages of 0 and 10, while the highest number of patients with transudates and exudates fell between the age groups of 50 and 70 years of age (29 and 31.7%). There were only a few effusions encountered in patients over 80 years of age (2.2%).

The etiological basis for transudates is recorded in table III. The highest number of the transudates was due to cardiac decompensation (65.6%). It was followed by transudates due to local stasis, such as liver cirrhosis (19.2%). Renal disease and hypoproteinemia accounted for only 14% of the transudates.

Table III. Types of transudates (n = 343)

	Number	Percentage
Due to cardiac decompensation	225	65.6
Due to local stasis	66	19.2
Due to renal disease	38	11.1
Due to hypoproteinemia	10	2.9
Due to Meigs' syndrome	4	1.2
Total	343	100

Table IV. Types of inflammatory exudates (n = 279)

Etiology	Number	Percentage
Purulent pericarditis	14	5
Rheumatic heart disease	18	6.5
Myocardial infarction	19	6.8
Pulmonary disease	16	5.7
Empyema	25	8.9
Pulmonary infarction	36	12.9
Tuberculosis	48	17.3
Unknown infections	10	3.6
Eosinophilic pleural effusions	7	2.5
Acute peritonitis	27	9.7
Tuberculous peritonitis	28	10
Diverticulitis	23	8.3
Subdiaphragmatic abscess	8	2.8

Table IV lists the inflammatory exudates. There were 279 inflammatory exudates amongst our specimens, of which 51 involved the pericardial cavity, 142 the pleural cavity, and 86 the peritoneal cavity. The most common inflammatory exudate of the pericardium was observed after myocardial infarction (6.8%), while purulent pericarditis (5%) accounted for the smallest number of pericardial effusions. No fungal infections or viral infections were encountered among the pericardial exudates.

The most common type of pleural exudate was observed in pulmonary tuberculosis (17.3%). We did not specify the type of pulmonary tuberculosis, but had the impression that fibrocaseous tuberculosis produced more frequently pleural exudates than the other forms of tuberculosis. The second most common cause of inflammatory exudates was pulmonary infarction (12.9%). The fluid was always hemorrhagic and contained sometimes clumps of clotted blood. Acute post-pneumonic infections leading to empyema accounted for 8.9% of inflammatory exudates, while chronic pulmonary disease, such as chronic bronchitis, bronchiectasis, or organizing pneumonia produced 5.7% of the inflammatory exudates. In 3.6%, the pleural exudates were due to unknown infections which might have been fungi. In seven cases, we encountered strong eosinophilic pleural effusions with eosinophils present from 25 to 60%. We successfully excluded the presence of parasitic infections, but could not explain the origin of the eosinophils.

Abdominal effusions in the form of ascites or peritoneal exudates were encountered in 30.8% of our cases. The most common cause was tuberculous peritonitis which we encountered in 10%. The least common cause was subdiaphragmatic abscess, which we encountered only in 2.8%.

In 430 effusions, or 40.2% of our material, we encountered malignant cells indicative of a primary or secondary malignant neoplasm (table VIII). We divided our cases into those in which malignant cells were unmistakably present and could be easily recognized. The second column lists those cases in which atypical cells were present which were strongly suspicious of malignancy, while a third column lists the cases in which no malignant cells were encountered, although the clinical and pathological findings strongly suggested the presence of a malignant neoplasm. In the evaluation of the accuracy of our material we had two choices, either to count the cases with definite malignant cells as positive and the cases with suspicious cells and negative findings as negative, or to add the cases with suspicious cells to the positive cases, as many others have done. We chose to count only those cases as positive in which unmistakably malignant cells were present. With those considerations in mind, table V shows 29 cases which were positive for malignant cells, while two cases had only suspicious cells, and two cases were negative, giving a total accuracy according to our rules of 87.8%. The interesting feature of table V was the fact that both cases of primary mesothelioma of the pericardium showed malignant cells.

Table VI analyzes the types of neoplastic exudates found in 240 cases of pleural exudates and transudates. Of a total number of 240 cases, 222 were positive for malignant cells. Five cases of carcinoma of the lung had only suspicious cells, while two were negative. Three cases of carcinoma of the breast had only suspicious cells, while one was negative. A total of 13 cases

Table V. Types of neoplastic exudates: pericardium

Diagnosis	Number of cases	Positive for malignant cells	Suspicious	Negative
Mesothelioma of the pericardium	2	2	0	0
Carcinoma of the lung	12	10	1	1
Carcinoma of the breast	9	8	0	1
Carcinoma of the thyroid	2	2	0	0
Lymphoma	6	5	1	0
Melanoma	2	2	0	0
Total number of cases	33	29	2	2
Percentage		87.8	6.1	6.1

Table VI. Types of neoplastic exudates: pleura

Diagnosis	Number of cases	Positive for malignant cells	Suspicious	Negative
Primary mesothelioma	17	17	0	0
Carcinoma of lung	61	54	5	2
Carcinoma of breast	52	48	3	1
Carcinoma of stomach	6	6	0	0
Carcinoma of colon	12	12	0	0
Carcinoma of pancreas	2	2	0	0
Carcinoma of liver	1	1	0	0
Carcinoma of uterus	5	4	1	0
Carcinoma of cervix	3	3	0	0
Carcinoma of testes	3	3	0	0
Carcinoma of kidney	12	10	2	0
Carcinoma of bladder	6	5	0	1
Carcinoma of prostate	6	5	1	0
Carcinoma of ovary	18	18	0	0
Melanoma	5	5	0	0
Lymphosarcoma	16	14	1	1
Reticulum cell sarcoma	4	4	0	0
Spindle cell sarcoma	6	6	0	0
Tumors of unknown origin	5	5	0	0
Total	240	222	13	5
Percentage		92.5	5.4	2.1

Table VII. Types of neoplastic exudates: peritoneum

Diagnosis	Number of cases	Positive for malignant cells	Suspicious	Negative
Primary mesothelioma	4	4	0	0
Carcinoma of ovary	18	17	1	0
Carcinoma of uterus	8	8	0	0
Carcinoma of cervix	14	14	0	0
Carcinoma of lungs	8	7	1	0
Carcinoma of breast	12	10	0	2
Carcinoma of kidney	6	6	0	0
Carcinoma of adrenal	1	1	0	0
Carcinoma of liver	2	2	0	0
Carcinoma of gallbladder	4	3	1	0
Carcinoma of pancreas	8	6	1	1
Carcinoma of stomach	10	8	2	0
Carcinoma of colon	21	19	1	1
Carcinoma of prostate	8	7	1	0
Carcinoma of bladder	4	4	0	0
Lymphosarcoma	9	8	1	0
Reticulum cell sarcoma	4	4	0	0
Hodgkin's disease	6	6	0	0
Melanoma	4	4	0	0
Spindle cell sarcoma	3	3	0	0
Tumors of unknown origin	3	2	1	0
Total	157	143	10	4
Percentage		91.1	6.4	2.5

in the pleural fluids had suspicious cells, while five cases had no malignant cells. This gives an accuracy of 92.5% if we count only positive cases. Especially remarkable are the five tumors of unknown origin which all were strongly positive, although it was impossible to recognize the origin of the tumors. The reason why cases of metastatic carcinoma of the lung showed the least accuracy in cancer cell detection was probably the great similarity between mesothelial cells and the small type of cells from bronchogenic carcinoma.

Table VII analyzes the types of neoplastic exudates found in the peritoneal cavity. Among the total number of 157 cases of metastatic tumors to

Table VIII. Summary of all neoplastic exudates

Diagnosis	Number of cases	Positive for malignant cells	Suspicious	Negative
Primary mesothelioma	23	23	0	0
Carcinoma of lung	81	71	7	3
Carcinoma of breast	73	66	3	4
Carcinoma of GI tract	66	59	5	2
Carcinoma of GU tract	45	40	4	1
Carcinoma of female organs	66	64	2	0
Other tumors	23	23	0	0
Blood dyscrasias	45	43	2	0
Tumors of unknown origins	8	7	1	0
Total	430	396	24	10
Percentage		92.1	5.6	2.3

the peritoneum, 143 or 91.1% were positive, 6.4% were suspicious and only 2.5% were negative. Again, metastatic carcinoma of the breast was missed more frequently, together with carcinoma of the stomach. In all three types of tumors, the glandular arrangement of the tumor itself is very loose and many metastatic cells propagate as single cell groups. The highest number of metastasizing malignant tumors in the peritoneal cavity was in carcinoma of the colon and carcinoma of the ovary. The latter type of tumor forms large clumps of adenocarcinoma with considerable mucous which makes it easy to recognize their origin.

Table VIII represents a summary of all malignant tumors which metastasized in the four serous cavities. Their total number was 430, with carcinoma of the lung and carcinoma of the breast leading the group followed by carcinoma of the gastrointestinal tract and carcinoma of the female sex organs. There was also an uncommonly large number of cells from metastatic blood dyscrasias such as Hodgkin's disease and lymphoma. 92.1% of the metastatic tumors exfoliated malignant cells in the serous cavities. In only 2.3% of our specimens were the cytological findings negative, while in 5.6% suspicious cells were present which, however, according to our principle, we did not count as positive malignant cells.

Discussion

Our first series of fluid examinations in 1962 included 754 exudates and transudates, of which 539 were pleural fluids and 215 ascites fluids. We examined 359 fluids of confirmed cases of cancer with involvement of the serous cavities and obtained positive results in 322, or 89.7%. Examination of the pleural fluids gave a higher percentage of positive results (91.2%) than the examination of ascites fluids (86%). We encountered a total of 37 false-negative cases, or 10.3%. 209 specimens were examined from patients who did not have malignant tumors, as proven by surgery or autopsy. The accuracy for confirmed negative cases was 90.9%. In 19 fluids, a false-positive diagnosis was made. In contrast to our other cytological examinations, in which the percentage of false-negatives was usually much higher than the percentage of false-positives, the error in diagnosis of fluid examinations was equal in the positive and negative groups (89.7 vs. 90.9%). This means that the distribution of cells played a less important role in the examination of tissue fluids than the recognition of cell types. Only ten unconfirmed positive tissue fluids were obtained (1.3%). The greatest difficulty in interpretation of the cytology of tissue fluids was the recognition of small and relatively undifferentiated malignant cells, such as cells from carcinoma of the lung or the breast from atypical mesothelial cells as they appeared under the influence of hypoxia.

The best procedure in the examination of tissue fluids seemed to be to examine first the mesothelial cells for their morphology and staining qualities.

After malignant cells have been recognized in pleural or ascites fluid, the important diagnostic problem of course remains to decide the type and site of the primary malignant tumor. In many cases, the existence of such a tumor is known and one has only to decide whether the malignant cells found in the fluid are compatible with a metastatic spread of this tumor. Much more difficult is the problem of deciding the type and site of an occult primary cancer which has metastasized to the pleura or peritoneum and exfoliated its cells into the tissue fluid. In these instances, it is often only possible to reach a tentative diagnosis on the basis of preference and give the clinician a lead where to search for the primary tumor. Such cases include melanomas and renal cell carcinomas. In order to be most effective in making such educated guesses, the cytopathologist must be fully familiar with the clinical history of the patient.

X. The Preparation of Specimens

In order to assure the best results in the cytological examination of transudates and exudates, an adequate technique must be employed for the collection and the preparation of the material. Aspiration of fluid from the serous cavities usually is a simple technical procedure and will deliver sufficient material for the examination of the cells. It is very important that smears are prepared as soon as possible after the fluid has been collected and that the cells are fixed as soon as possible in order to prevent autolytic changes. An 18-gauge needle with an obturator and a 25-ml syringe are the most practical instruments for the aspiration of exudates and transudates. From the aspirated material, slides are prepared which are immediately fixed. To prevent the cells from washing off in the fixing fluid, a thin coat of egg albumin should be spread over the slides before the cell suspension is added. This is not necessary when the cells are suspended in mucous or similar protein-rich material.

Modern cytology now has three methods available for the cytological approach to clinical problems [138]: those based upon exfoliative cytology, those based upon abrasive cytology, and those based upon aspiration cytology. The methods of exfoliative cytology and abrasive cytology examine cells derived from the surfaces of mucous membranes and serous membranes. Aspiration cytology examines cells obtained by negative pressure from solid tissues. Normally, serous cavities contain a small amount of colorless fluid containing some mucoproteins. A decrease of the osmotic pressure of the blood or an increase of capillary permeability may cause an accumulation of excessive amounts of those fluids known as transudates. An increased hydrostatic pressure of the blood is easily caused by an increased venous pressure that may be general or local. Transudates will usually recur after their removal unless the causal factor has been removed. Exudates are produced as the result of inflammatory conditions which include infectious processes and invasion of the serous cavities by tumor cells.

Samples of transudates and exudates can usually be easily obtained by paracentesis under local anesthesia. To prevent coagulation of the aspirated fluid, it is recommended that 10 ml of a 2.5% sterile sodium citrate solution should be added to each 100 ml of effusion. The fluid should be collected in sterile containers and more than one specimen vial should be sent to the laboratory in order that various test may be performed simultaneously.

Fluid from the pericardial cavity is obtained by inserting the needle in the fifth intercostal space below the left nipple with the shaft of the needle directed upward and medially. If the needle touches the heart, a pulse-like movement of the instrument can be observed. Fluid from the pleural cavity is obtained by thoracentesis. The patient sits in bed leaning forward with his arms raised. The area of greatest dullness is determined by percussion. The puncture should be in the mid-axillary line and close to the upper edge of the ribs. The fluid is best obtained by aspiration with a 16-gauge needle into a syringe or aspiration into an aspirator bottle.

Fluid from the abdominal cavity is obtained by paracentesis of the abdomen. The best point for the abdominal puncture is the midline about halfway between the symphysis pubis and the umbilicus. If a large amount of fluid is to be withdrawn, it is advisable to withdraw the fluid slowly.

It is always recommended to study the physical characteristics of the fluid before it is sent to the laboratory. The volume of transudates and exudates depends greatly upon the duration of the disease and the space available for the accumulation of fluid. In the abdomen, the volume may reach 15-25 liters, while in the chest transudates rarely exceed 3 liters in each pleural cavity, and in the pericardial cavity 600 ml appears to be the maximum of pericardial fluid. The color of the fluid is of considerable importance. Transudates are usually colorless or possess a slight yellow hue. A mixture of red cells may give the fluid a reddish tinge. A mixture of white blood cells and tumor cells will produce a white to yellow color, while chyle will give the exudate a milky appearance. The turbidity of transudates and exudates depends upon the amount of suspended formed particles. All degrees of slight cloudiness to thick creamy pus may be observed. The coagulability of transudates and exudates depends upon the amount of fibrin present in the fluid. Protein-rich fluids from metastatic carcinoma can form a heavy coagulant. Coagulates appear very quickly after removal of the fluid from the serous cavity and for this reason the addition of sterile sodium citrate to any tissue fluid is strongly recommended.

The specific gravity of fluids obtained by paracentesis is of utmost importance in differentiating between transudates and exudates. With sufficient quantities available, a simple Squibb urinometer may be used. Transudates usually show a specific gravity between 1,006 and 1,018, whereas exudates, due to their greater content of protein, have a specific gravity of over 1,020. By carefully checking on those physical qualities, it will be usually possible to classify accurately the type of the fluid and will give us important leads as to additional tests to be employed for the accurate diagnosis of the causal factors involved.

The further examination of exudates and transudates can be divided into chemical examinations, bacteriological examinations, serological examinations, and cytological examinations. Since this monograph is devoted to the cytology of exudates and transudates, we shall only discuss techniques used for this one type of examination and refer our readers to textbooks on clinical pathology for the other methods.

Immediate fixation is an important step for the preparation of good cytologic specimens. The fixative specifically recommended for cytological specimens is alcohol, which causes cells to shrink with coarsening of the cellular structures and sharpening of nuclear chromatin patterns. Originally, a mixture of 95% ethyl alcohol and ether in an equal proportion was recommended, but now most cytopathology laboratories use only 95% alcohol. Methanol produces less shrinkage than ethanol, but preserves delicate chromosomal details. All the fixatives can be applied as wet fixatives and the slides should remain submerged in the fixing fluid for a minimum of 15–30 min. Wet fixatives are sometimes replaced by coating fixatives which are a combination of alcohol and a waxlike substance which forms a thin protective coating over the cells [27]. They can be applied as aerosols or in a liquid base. Coating fixative must be removed from the cell samples before staining, which can be done by immersing the slides for 10 min in water. An inexpensive coating fixative are hairsprays which have a high alcohol content.

If the cell sample cannot be brought to the laboratory within 12–24 h, a prefixation of 50% ethanol may be necessary. If cells should be prepared for cell block preparations, 10% formalin should be used as fixative. Buffered 10% formalin is a good preservative for lipids, carbohydrates, and mucin. Alcohol fixation results in faster fixation, better preservation of glycogen, but the loss of fat and lipids.

Many staining methods are available. The most popular one in the United States is the one recommended by PAPANICOLAOU, which in our laboratory has always given excellent results. The staining method which we use has been taken from our chapter on clinical cytology which appeared in *A Textbook of Clinical Pathology* by MILLER [86]. It is reprinted below and consists of 24 steps: (1) fix slides for 15–30 min in 95% alcohol; (2) 1 min in 70% alcohol; (3) 1 min in 50% alcohol; (4) 2 min in distilled water;

(5) 5 min in Harris hematoxylin; (6) 1 min in distilled water; (7) two quick dips in 0.2% hydrochlorid acid; (8) 2-min rinse in running tap water; (9) 2 min in lithium carbonate (1 g in 1,000 ml distilled H_2O); (10) 3 min in running tap water; (11) 1 min in 70% alcohol; (12) 1 min in 95% alcohol; (13) 5 min in OG-6; (14) 1 min in 95% alcohol; (15) 1 min in 95% alcohol; (16) 5 min in EA-50; (17) 1 min in 95% alcohol; (18) 1 min in 95% alcohol; (19) 2 min in absolute alcohol; (20) 2 min in absolute alcohol; (21) 2 min in an equal mixture of absolute alcohol and xylene; (22) 2 min in xylene; (23) 5 min in xylene, and (24) coverslip using Permount (do not let slide dry). Liquid plastic Diatex may also be used instead of glass coverslips.

The staining is performed in a set of glass jars containing metal slide holders which permit the slides to be passed from one solution into the other. The time the slides remain in each solution can be measured by minutes with a stopwatch or by dips, 1 min corresponding roughly to 15 3-sec dips. Each staining tray must be well-drained and each bottle blotted in order to prevent the spoiling of the stains. Frequent gentle agitation for each solution is essential.

While the Papanicolaou stain is preferred in most American cytological laboratories, stains using hematoxylin-eosin stain, the Giemsa or Wright stain or the Shorr stain have been used with equal success. Cytochemical stains are recommended for cytochemical analysis, such as the Feulgen stain for the demonstration of DNA and methyl green-pyronin for the demonstration of RNA. Newer techniques include the acridine orange stains, fluorescent antibody techniques using the fluorescent microscope, and examination of all smears with phase microscope and the electron microscope. However, at the present time all these sophisticated methods possess little practical diagnostic value.

References

1 ALFREY, A.C.; GOSS, J.E., and OGDEN, D.A.: Uremic hemopericardium. Am. J. Med. *45:* 391–400 (1968).

2 ALP, C. and KARADENIZ, A.: Four cases of malignant mesothelioma of the pleura. Int. Surg. *47:* 150–158 (1967).

3 ALTEMEIER, W.A.; CULBERTSON, W.R.; FULLEN, W.D., and SHOOK, C.D.: Intra-abdominal abscesses. Am. J. Surg. *125:* 70–79 (1973).

4 ARIEL, I.M.; OROPEZA, R., and PACK, G.T.: Intracavitary administration of radioactive isotopes in the control of effusions due to cancer. Results in 267 patients. Cancer *19:* 1096–1102 (1966).

5 ARIEL, I.M.; OROPEZA, R., and PACK, G.T.: The treatment of cancerous effusion. Prog. clin. Cancer *2:* 185–192 (1966).

6 BACKER, G. DE; HOVE, W. VAN et PANNIER, R.: Tamponnade cardiaque: première manifestation d'une tumeur maligne. Cœur Méd. Interne *11:* 411–418 (1972).

7 BACKMAN, A. and PASILA, M.: Pleural biopsy in the diagnosis of pleural effusion. Scand. J. resp. Dis. Suppl. *89:* 155–157 (1974).

8 BAKALOS, D.; CONSTANTAKIS, N., and TSICRICAS, T.: Distinction of mononuclear macrophages from mesothelial cells in pleural and peritoneal effusions. Acta cytol. *18:* 20–22 (1974).

9 BAKALOS, D.; CONSTANTAKIS, N., and TSICRICAS, T.: Recognition of malignant cells in pleural and peritoneal effusions. Acta cytol. *18:* 118–121 (1974).

10 BARBEE, C.L. and GILSDORF, R.B.: Diagnostic peritoneal lavage in evaluating acute abdominal pain. Ann. Surg. *181:* 853–856 (1975).

11 BENEDICT, W.F.; BROWN, C.D., and PORTER, I.H.: Long acrocentric marker chromosomes in malignant effusions and solid tumors. N.Y. J. Med. *9:* 952–955 (1971).

12 BENEDICT, W.F. and PORTER, I.H.: The cytogenetic diagnosis of malignancy in effusions. Acta cytol. *16:* 304–306 (1972).

13 BERGE, T. and HELLSTEN, S.: Cytological diagnosis of cancer cells in pleural and ascitic fluid. Comparison of results obtained by pathologist and cytologist. Acta cytol. *10:* 138–140 (1966).

14 BERGER, H.W. and SECKLER, S.G.: Pleural and pericardial effusions in rheumatoid disease. Ann. intern. Med. *64:* 1291–1297 (1966).

15 BERNARD, S.; VERAN, P. et CAM, M. LE: Etudes sur l'équipement enzymatique des épanchements pleuro-péritonéaux. Rapports entre les activités de quelques systèmes enzymatiques dans ces liquides d'épanchement et dans le sérum sanguin. Annls Biol. clin. *24:* 679–688 (1966).

16 BERSON, D.: Letter: Pleural lavage in carcinoma of the esophagus. New Engl. J. Med. *291:* 530 (1974).

17 BLACK, L.F.: The pleural space and pleural fluid. Mayo Clin. Proc. *47:* 493–506 (1972).

18 BODDINGTON, M.M.; SPRIGGS, A.I.; MORTON, J.A., and MOWAT, A.G.: Cytodiagnosis of rheumatoid pleural effusions. J. clin. Path. *24:* 95–106 (1971).

19 BOSCARO, M. e CAPRIOGLIO, A.: La puntura del Douglas nella citodiagnostica delle neoplasie ovariche. Minerva med., Roma *60:* 1332–1339 (1969).

20 BOWER, G.: Eosinophilic pleural effusion. A condition with multiple causes. Am. Rev. resp. Dis. *95:* 746–751 (1967).

21 BRANDT, H.J. *et al.:* Differential diagnosis of pleural effusion using thoracoscopy. Pneumonologie *145:* 192–203 (1971).

22 BROWN, A.K.: Chronic idiopathic pericardial effusion. Br. Heart J. *28:* 609–614 (1966).

23 BROWN, J.W.; KRISTENSEN, K.A., and MONROE, L.S.: Peritoneal mesothelioma following pneumoperitoneum maintained for twelve years. Report of a case. Am. J. dig. Dis. *13:* 830–835 (1968).

24 BRUNO, M.S. and OBER, W.B.: Recurrent chylous ascites. N.Y. J. Med. *70:* 282–290 (1970).

25 BUJA, L.M.; FRIEDMAN, C.A., and ROBERTS, W.C.: Hemorrhagic pericarditis in uremia. Clinicopathologic studies in six patients. Archs Path. *90:* 325–330 (1970).

26 BURGERMAN, A.: Myxedema with pericardial effusion. Med. Ann. Distr. Columbia *35:* 126–133 (1966).

27 BURROWS, S.; CARPENTER, E., and WARREN, L.S.: Centrifuged coverglass method for cytologic study of body fluids. Acta cytol. *12:* 404–405 (1968).

28 CAILLEAU, R.; YOUNG, R.; OLIVÉ, M., and REEVES, W.J., jr.: Breast tumor cell lines from pleural effusions. J. natn. Cancer Inst. *53:* 661–674 (1974).

29 CALLE, S.: Megakaryocytes in an abdominal fluid. Acta cytol. *1:* 78–80 (1968).

30 CAMERON, J.L.; ANDERSON, R.P., and ZUIDEMA, G.D.: Pancreatic ascites. Surgery Gynec. Obstet. *125:* 328–332 (1967).

31 CAMPBELL, R.A.: Pneumococcal peritonitis in cirrhosis. New Engl. J. Med. *278:* 743 (1968).

32 CARDOZO, P.L.: A critical evaluation of 3,000 cytologic analyses of pleural fluid, ascitic fluid and pericardial fluid. Acta cytol. *6:* 455–460 (1966).

33 CARDOZO, P.L.: Letters to the Editors: Cytology of effusions. Acta cytol. *2:* 85–86 (1968).

34 CARDOZO, P.L.: Malignitätskriterien in Perikard- und Pleuraergüssen sowie in Acites. Med. Lab. *23:* 139–142 (1970).

35 CARDOZO, P.L. and HARTING, M.C.: On the function of lymphocytes in malignant effusions. Acta cytol. *4:* 307–313 (1972).

36 CARR, D.T.: Diagnostic studies of pleural fluid. Surg. Clin. N. Am. *53:* 801–804 (1973).

37 CASTOR, C.W. and NAYLOR, B.: Characteristics of normal and malignant human mesothelial cells studied *in vitro*. Lab. Invest. *20:* 437–443 (1969).

38 CEELEN, G.H.: The cytologic diagnosis of ascitic fluid. Acta cytol. *8:* 175–185 (1964).

39 CHAM, W.C. *et al.:* Radiation therapy of cardiac and pericardial metastases. Radiology *114:* 701–704 (1975).

40 CLARKSON, B.; OTA, K.; OHKITA, T., and O'CONNOR, A.: Kinetics of proliferation of cancer cells in neoplastic effusions in man. Cancer *18:* 1189–1213 (1965).

41 CONTINO, C.A. and VANCE, J.W.: Eosinophilic pleural effusion. N.Y. J. Med. *66:* 2044–2048 (1966).

42 COUPLAND, G.: Pancreatic ascites in childhood. J. pediat. Surg. *5:* 570 (1970).

43 CRAIG, R.; SPARBERG, M.; IVANOVICH, P., and RICE, L.: Nephrogenic ascites. Archs intern. Med. *134:* 276–279 (1974).

44 CURRAN, W.S.; WILSON, J.N., and SHOOP, J.D.: Eosinophilic pleural effusion and bronchiectasis. Rocky Mountain med. J. *65:* 49–53 (1968).

45 DALQUEN, P.; DABBERT, A.F. und HINZ, I.: Zur Epidemiologie der Pleuramesotheliome. Vorläufiger Bericht über 119 Fälle aus dem Hamburger Raum. Prax. Pneumol. *23:* 547–558 (1969).

46 DARNIS, F.: Review. Cirrhosis edema-ascites syndrome. The role of hormonal abnormalities in the pathogenesis. Minn. Med. *54:* 143–150 (1971).

47 DAVIS, J.H.: Current concepts of peritonitis. Am. Surg. *33:* 673–681 (1967).

48 DAVIS, J.M.: Ultrastructure of human mesotheliomas. J. natn. Cancer Inst. *52:* 1715–1725 (1974).

49 DAVIS, P.J. and JACOBSON, S.: Myxedema with cardiac tamponade and pericardial effusion of 'gold paint' appearance. Archs intern. Med. *120:* 615–619 (1967).

50 DAVIS, R.H. and McGOWAN, L.: Comparative peritoneal cellular content as related to species and sex. Anat. Rec. *162:* 357–361 (1968).

51 DEBRUX, J.A.; DUPRÉ-FROMENT, J., and MINTZ, M.: Cytology of the peritonale fluids sampled by coelioscopy or by cul-de-sac puncture. Its value in gynecology. Acta cytol. *12:* 395–403 (1968).

52 DEGEORGES, M.; LEBLANC, P. et MORIN, B.: Epanchements péricardiques de l'insuffisance thyroïdienne. A propos de trois observations. Sem. Hop. Paris *41:* 2704–2714 (1965).

53 DEKKER, A.; GRAHAM, T., and BUPP, P.A.: The occurrence of sickle cells in pleural fluid. Report of a patient with sickle cell disease. Acta cytol. *19:* 251–254 (1975).

54 DE LA MAZA, L.M.; THAYER, B.A., and NAEIM, F.: Cylindroma of the uterine cervix with peritoneal metastases: report of a case and review of the literature. Am. J. Obstet. Gynec. *112:* 121–125 (1972).

55 DEMPSEY, J.J.; EISSA, A.; ATTIA, M., and RAMZY, A.: Pericardial effusion of 'gold-paint' appearance following myocardial infarction. Archs intern. Med. *118:* 249–254 (1966).

56 DOMAGALA, W. and WOYKE, S.: Transmission and scanning electron microscopic studies of cells in effusions. Acta cytol. *19:* 214–224 (1975).

57 DONOWITZ, M.; KERSTEIN, M.D., and SPIRO, H.M.: Pancreatic ascites. Medicine *53:* 183–195 (1974).

58 DUPRÉ-FROMENT, J.; MINTZ, M. et BRUX, J. DE: Cytologie péritonéale par prélèvements per-coelioscopiques ou par ponction du Douglas. Gynéc. Obstét. *66:* 95–108 (1967).

59 Editorial: Pleural effusion. Br. med. J. *iii:* 192–193 (1975).

60 EPSTEIN, M.; CALIA, F.M., and GABUZDA, G.J.: Pneumococcal peritonitis in patients with postnecrotic cirrhosis. New Engl. J. Med. *278:* 69–73 (1968).

61 ESLAMI, B. and LUTCHER, C.L.: Antemortem diagnosis in 2 cases of malignant peritoneal mesothelioma. Am. J. med. Sci. *267:* 117–121 (1974).

62 EVANS, C.; RASHID, A.; ROSENBERG, I.L., and POLLOCK, A.V.: An appraisal of peritoneal lavage in the diagnosis of the acute abdomen. Br. J. Surg. *62:* 119–120 (1975).

63 EXADAKTYLOS, P.: Zur Zytologie von Transsudaten. 1. Mitteilung. Die Zytomorpho-
logie pleuraler Transsudate. *101:* 729–736 (1974).

64 FARAH, M.G.; NASSAR, V.H., and SHAHID, M.: Marked eosinophilia and eosinophilic pleural effusion in Hodgkin's disease. Report of a case with review of literature.
J. Med. Liban *26:* 513–521 (1973).

65 FAWAL, I.A.; KIRKLAND, L.; DYKES, R., and FOSTER, G.L.: Chronic primary chylopericardium. Report of a case and review of the literature. Circulation *35:* 777–782
(1967).

66 FEIZI, O.; GRUBB, C.; SKINNER, J.I.; CONSTANTINIDOU, M., and HENDERSON, W.G.:
Primary atypical pneumonia due to mycoplasia pneumoniae complicated by haemorrhagic pleural effusion, haemolytic anaemia and myocarditis. Br. J. clin. Pract. *27:*
99–101 (1973).

67 FLANNERY, E.P.; GREGORATOS, G., and CORDER, M.P.: Pericardial effusions in patients with malignant diseases. Archs intern. Med. *135:* 976–977 (1975).

68 FOOT, N.C.: The identification of neoplastic cells in serous effusions. Critical analysis
of smears from 2029 persons. Am. J. Path. *30:* 661 (1956).

69 FOWLER, R.: Primary peritonitis: changing aspects 1956–1970. Aust. Paediat. J. *7:*
73–83 (1971).

70 FRANCHI, F.; NAVA, C. e CAVAZZUTI, F.: Contributo alla citologia dei rersamenti
delle sierose (esame di 838 liquidi). Omnia Med. *45:* 275–297 (1967).

71 FRANCO, A.E.; LEVINE, H.D., and HALL, A.P.: Rheumatoid pericarditis. Report of
17 cases diagnosed clinically. Ann. intern. Med. *77:* 837–844 (1972).

72 FRENI, S.C.; JAMES, J., and PROP, F.J.: Tumor diagnosis in pleural and ascitic
effusions based on DNA cytophotometry. Acta cytol. *15:* 154–162 (1971).

73 FRIEDMAN, B.J. and SEGAL, B.L.: Chronic effusive pericarditis associated with healed
myocardial infarction. Report of a case. Dis. Chest *49:* 217–221 (1966).

74 FROST, J.K.: The cell in health and disease, pp. 101–122 (Karger, Basel 1969).

75 GAENSLER, E.A.: 'Idiopathic' pleural effusion. New Engl. J. Med. *283:* 816–817 (1970).

76 GAENSLER, E.A. and KAPLAN, A.I.: Asbestos pleural effusion. Ann. intern. Med. *74:*
178–191 (1971).

77 GALY, P.; BRUNE, J. et ROUMAGOUX, J.: La pleurésie à éosinophiles autonomes. J.
fr. Méd. Chir. thorac. *24:* 25–34 (1970).

78 GALY, P.; BRUNE, J. et DORSIT, G.: Etude statistique de 710 épanchements pleuraux
observés dans un service de pneumologie. Lyon méd. *226:* 279–285 (1971).

79 GIRARD, M.; LOMBARD-PLATET, R. et MOULINIER, B.: Syndrome ascitique appa-
remment isolé: premier signe d'une pancréatite subaiguë. J. Méd. Lyon *46:* 1015–
1021 (1965).

80 GOVAERTS, J.P.; KAHN, R.J.; VEREERSTRAETEN, P.; PRIMO, G. et TOUSSAINT, C.: La
péricardite hémorragique du mal de Bright. A propos de quatre observations. Acta
cardiol. *25:* 58–68 (1970).

81 GRAHAM, J.; BURSTEIN, P., and GRAHAM, R.: Prognostic significance of pleural
effusion in ovarian cancer. Am. J. Obstet. Gynec. *106:* 312–313 (1970).

82 GRAHAM, R.M.; BARTELS, J.D., and GRAHAM, J.B.: Screening for ovarian cancer by
cul-de-sac aspiration. Acta cytol. *6:* 492–495 (1962).

83 GRUENZE, H.: Klinische Zytologie der Thoraxkrankheiten, pp. 70–96 (Enke, Stuttgart 1955).

84 HAAM, E. VON: Fundamentals of disease, pp. 235–237 (Springer, New York 1953).

85 HAAM, E. VON: Clinical cytology; in MILLER A textbook of clinical pathology, pp. 831–881 (Williams & Wilkins, Baltimore 1966).

86 HAAM, E. VON: Clinical cytology; in MILLER and WELLER Textbook of clinical pathology, pp. 225–229 (Williams & Wilkins, Baltimore 1971).

87 HAAM, E. VON: Radiation cell changes; in WEID, KOSS and REAGAN Compendium on diagnostic cytology, pp. 242–266 (Tutorials of Cytology, Chicago 1976).

88 HAIN, E. und ENGEL, J.: Zur Diagnose, Differentialdiagnose und Epidemiologie der Pleuraergüsse. Pneumonologie 145: 175–184 (1971).

89 HANCOCK, E.W.: Subacute effusive-constrictive pericarditis. Circulation 43: 183–192 (1971).

90 HERTZOG, P.; TOTY, L. et PERSONNE CHEVASSU, C.: Tumeurs d'évolution maligne sur sequettes pleurales importantes (à propos de quatre observations). J. fr. Méd. Chir. thorac. 22: 339–353 (1968).

91 HIGBY, D.J. and OHNUMA, T.: Plasmacytoma cell ascites. N.Y. St. J. Med. 75: 1074–1076 (1975).

92 HIRSCH, D.M., jr.; NYDICK, I.N., and FARROW, J.H.: Malignant pericardial effusion secondary to metastatic breast carcinoma. A case of long-term remission. Cancer 19: 1269–1272 (1966).

93 HOFF, D.D. and LIVOLSI, V.: Diagnostic reliability of needle biopsy of the parietal pleura. A review of 272 biopsies. Am. J. clin. Path. 64: 200–203 (1975).

94 HOLLANDER, J.L.; McCARTY, D.J., jr.; ASTORGA, G., and CASTRO-MURILLO, E.: Studies on the pathogenesis of rheumatoid joint inflammation. Ann. intern. Med. 62: 271 (1965).

95 HUDSPETH, A.S. and MILLER, H.S.: Isolated (primary) chylopericardium. Diagnosis and surgical treatment. J. thorac. cardiovasc. Surg. 51: 528–531 (1966).

96 HUNT, C.E.; PAPERMASTER, T.C.; NELSON, E.H., and KRIVIT, W.: Eosinophilic peritonitis. Report of two cases. Lancet 87: 473–476 (1967).

97 JACOBS, D.S. and REYNARD, J.D.: Chylangioma of the mesentery: content and pathogenesis. J. Kans. med. Soc. 73: 145–147 (1972).

98 JAHODA, E.; BARTELS, P.H.; BIBBO, M.; BAHR, G.F.; HOLZNER, J.H., and WIED, G.L.: Computer discrimination of cells in serous effusions. I. Pleural fluids. Acta cytol. 17: 94–105 (1973).

99 JAHODA, E.; BARTELS, P.H.; BIBBO, M.; BAHR, G.F.; HOLZNER, J.H., and WIED, G.L.: Computer discrimination of cells in serous effusions. II. Peritoneal fluid. Acta cytol. 6: 533–537 (1973).

100 JARVI, O.H.; KUNNAS, R.J.; LAITIO, M.T., and TYRKKO, D.E.S.: The accuracy and significance of cytologic cancer diagnosis of pleural effusions. Acta cytol. 2: 152–158 (1972).

101 JOHNSON, W.D.: The cytological diagnosis of cancer in serous effusions. Acta cytol. 3: 161–173 (1966).

102 JONES, F.L., jr.: Subcutaneous implantation of cancer: a rare complication of pleural biopsy. Chest 57: 189–190 (1970).

103 KELLEY, S.; McGARRY, P., and HUTSON, Y.: Atypical cells in pleural fluid characteristic of systemic lupus erythematosus. Acta cytol. 15: 357–362 (1971).

104 KERN, W.H.: Cytology of effusions. Acta cytol. 11: 167 (1967).

105 KERN, W.H.: Benign papillary structures with psammoma bodies in culdocentesis fluid. Acta cytol. *13:* 178–180 (1969).

106 KERR, A., jr.: Myocardopathy, alcohol, and pericardial effusion. Archs intern. Med. *119:* 617–619 (1967).

107 KINDRED, L.H.; HEILBRUNN, A., and DUNN, M.: Cholesterol pericarditis associated with rheumatoid arthritis. Treatment of pericardiectomy. Am. J. Cardiol. *23:* 464–468 (1969).

108 KIRIANOFF, T.; STRAUCH, G.O., and ROGERS, J.F.: The nature of pancreatic ascites. Conn. Med. *33:* 573–575 (1969).

109 KNIGHT, H.F.: Primary chylopericardium. J. thorac. cardiovasc. Surg. *50:* 567–570 (1965).

110 KONIKOV, V.; BLEISCH, V., and PISKIE, V.: Prognostic significance of cytologic diagnoses of effusions. Acta cytol. *10:* 335–339 (1966).

111 KORNREICH, F. *et al.:* Pericardial disorders in neoplastic diseases. Acta cardiol. *22:* 426–443 (1967).

112 KOSS, L.G.: Diagnostic cytology and its histopathologic bases, pp. 490–537 (Lippincott, Philadelphia 1968).

113 LABAY, G.R. and FEINER, F.S.: Malignant pleural endometriosis. Am. J. Obstet. Gynec. *110:* 478–480 (1971).

114 LATHAM, B.A.: Pericarditis associated with rheumatoid arthritis. Ann. rheum. Dis. *25:* 235–241 (1966).

115 LEGRAND, M.; PARIENTE, R.; ANDRÉ, J.; CHRÉTIEN, J. et BROUET, G.: Ultrastructure de la plèvre pariétale humaine. Presse méd. *79:* 2515–2520 (1971).

116 LEGRAND, M.; ANDRÉ, J. et PARIENTE, R.: Ultrastructure de la cellule pleurale mésothéliale. Poumon Cœur *28:* 33–38 (1972).

117 LIGHT, R.W.; MACGREGOR, M.I.; LUCHSINGER, P.C., and BALL, W.C., jr.: Pleural effusions: the diagnostic separation of transudates and exudates. Ann. intern. Med. *77:* 507–513 (1972).

118 LIGHT, R.W.; EROZAN, Y.S., and BALL, W.C., jr.: Cells in pleural fluid. Their value in differential diagnosis. Archs intern. Med. *132:* 854–860 (1973).

119 LINDSAY, J., jr.; CRAWLEY, I.S., and CALLAWAY, G.M., jr.: Chronic constrictive pericarditis following uremic hemopericardium. Am. Heart J. *79:* 390–395 (1970).

120 LOIRE, R.; FROMENT, R.; GONIN, A. et PERRIN, A.: La biopsie du péricarde. Intérêt diagnostique dans les péricardites subaiguës et chroniques, à propos de 70 cas. Archs Mal. Cœur *68:* 11–17 (1975).

121 MAATHUIS, J.B.: Investigation of peritoneal fluid obtained by laparoscopy from gynecological patients. J. Reprod. Med. *12:* 248–249 (1974).

122 MALAVE, G.; FOSTER, E.D., and WILSON, J.A.: Bronchopleural fistula – present-day study of an old problem. A review of 52 cases. Ann. thorac. Surg. *11:* 1–10 (1971).

123 MARSAN, C.; MINET, F. et ROUJEAU, J.: Les épanchements à cellules mésothéliales. Confrontation anatomo-cytologique. Sem. Hôp. Paris *47:* 5595 (1971).

124 MASIN, F. and MASIN, M.: Cytodiagnosis of effusions by the desorption technic. Acta cytol. *9:* 380–385 (1965).

125 MASLAND, D.S.; ROTZ, C.T., and HARRIS, J.H., jr.: Postradiation pericarditis with chronic pericardial effusion. Ann. intern. Med. *69:* 97–102 (1968).

126 MATSUI, E. and ITO, T.: Marfan's syndrome complicated by huge hydropericardium. Jap. Heart J. *8:* 98–104 (1967).

127 MATZEL, W.: Zur funktionellen Beurteilung zytologischer Befunde von Pleura-
     höhlenergüssen. Z. Erkr. Atmungsorgane *132:* 293–301 (1970).
128 MAVROMATIS, F.: Über die Anwendung der PAS-Reaktion zur Identifizierung von
     Mesothelialzellen (Zusammenstellung und kritische Betrachtung der Ergebnisse).
     Acta histochem. *21:* 201–212 (1965).
129 MAVROMATIS, F.S.: Some morphologic features of cells containing PAS-positive
     intracytoplasmic granules in smears of serous effusions. Acta cytol. *8:* 426–430 (1964).
130 McDONALD, A.D. and CORBETT McDONALD, J.: Epidemiologic surveillance of meso-
     thelioma in Canada. Can. med. Ass. J. *109:* 359–362 (1973).
131 McDONALD, A.D.; MAGNER, D., and EYSSEN, G.: Primary malignant mesothelial
     tumors in Canada, 1960–1968. A pathologic review by the Mesothelioma Panel of
     the Canadian Tumor Reference Centre. Cancer *31:* 869–876 (1973).
132 McGOWAN, L.; DAVIS, R.H.; STEIN, D.B.; BEBON, S., and VASKELIS, P.: Cytologic
     differential of pelvic cavity aspiration specimens in normal women. Obstet. Gynec.,
     N.Y. *30:* 821–829 (1967).
133 McGOWAN, L. and DAVIS, R.H.: Cytology of serous fluid in the pelvic peritoneal
     cavity of pregnant and postpartum women. Am. J. clin. Path. *51:* 150–153 (1969).
134 McGOWAN, L. and DAVIS, R.H.: Peritoneal fluid cellular pattern in obstetrics and
     gynecology. Am. J. Obstet. Gynec. *106:* 979–995 (1970).
135 McGOWAN, L.; BUNNAG, B., and ARIAS, L.B.: Peritoneal fluid cytology associated
     with benign neoplastic ovarian tumors in women. Am. J. Obstet. Gynec. *113:*
     961–966 (1972).
136 McGOWAN, L. and BUNNAG, B.: A morphologic classification of peritoneal fluid
     cytology in women. Int. J. Gynaec. Obstet. *11:* 173–183 (1973).
137 McGOWAN, L. and BUNNAG, B.: Morphology of mesothelial cells in peritoneal fluid
     from normal women. Acta cytol. *3:* 205–209 (1974).
138 McGREW, E.A. and NANOS, S.: The cytology of serous effusions; in KEEBLER and
     REAGAN A manual of cytotechnology, pp. 197–207 (American Society of Clinical
     Pathologists, Chicago 1975).
139 McWHORTER, J.E., 4th and LeROY, E.C.: Pericardial disease in scleroderma (sys-
     temic sclerosis). Am. J. Med. *57:* 566–575 (1974).
140 MELAMED, M.R.: The cytological presentation of malignant lymphomas and related
     diseases in effusions. Cancer *16:* 413 (1963).
141 MEYERS, M.A.: The spread and localization of acute intraperitoneal effusions.
     Radiology *95:* 547–554 (1970).
142 MILLER, A.J. *et al.:* The production of acute pericardial effusion. The effects of
     various degrees of interference with venous blood and lymph drainage from the heart
     muscle in the dog. Am. J. Cardiol. *28:* 463–466 (1971).
143 MILNE, J.: Fifteen cases of pleural mesothelioma associated with occupational
     exposure to asbestos in Victoria. Med. J. Aust. *ii:* 669–673 (1969).
144 MIRIAJANIAN, A.; AMBRUOSO, V.N.; DERBY, B.M., and TICE, D.A.: Massive bilateral
     hemorrhagic pleural effusions in chronic relapsing pancreatitis. Archs Surg., Chicago
     98: 62–66 (1969).
145 MODAI, J.; HAZARD, J. et DOMART, A.: Epanchements pleuro-péritonéaux révéla-
     teurs de pancréatites chroniques. Intérêt du dosage de l'amylase dans les épanche-
     ments séreux. (A propos de trois observations.) Bull. Soc. Méd. Hôp. Paris *116:*
     1217–1230 (1965).

146 MOERTEL, C.G.: Peritoneal mesothelioma. Gastroenterology *63:* 346–350 (1972).

147 MOHANDAS, D.; SARATCHANDRA, R., and BHASKAR, A.G.: A case of pseudomyxoma peritonei. Am. J. Proctol. *22:* 44–46 (1971).

148 MONIF, G.R. and DALY, J.W.: 'Living cytology' in the diagnosis of intraabdominal adenocarcinoma. Obstet. Gynec., N.Y. *46:* 80–83 (1975).

149 MORRIS, P.J.: Diagnostic paracentesis of the acute abdomen. Br. J. Surg. *53:* 707–708 (1966).

150 MURAD, T.M.: Electron microscopic studies of cells in pleural and peritoneal effusions. Acta cytol. *17:* 401–409 (1973).

151 MURPHY, W.M. and NG, A.B.P.: Determination of primary site by examination of cancer cells in body fluids. Am. J. clin. Path. *58:* 479–488 (1972).

152 NATELSON, E.A.; ALLEN, T.W.; RIGGS, S., and FRED, H.L.: Bloody ascites: diagnostic implications. Am. J. Gastroent. *52:* 523–527 (1969).

153 OELS, H.C.; HARRISON, E.G., jr.; CARR, D.T., and BERNATZ, P.E.: Diffuse malignant mesothelioma of the pleura: a review of 37 cases. Chest *60:* 564–570 (1971).

154 PALIARD, P.; VINCENT, P. et BARTHE, J.: Place de la fluorescence à l'orange d'acridine dans la cytologie des épanchements des séreuses. Lyon méd. *214:* 488–492 (1965).

155 PAPOWITZ, A.J. and LI, J.K.: Abdominal sarcoidosis with ascites. Chest *59:* 692–695 (1971).

156 PEARLMAN, D.M.: Hepatogenic ascites: pathogenesis and management. N.Y. J. Med. *68:* 1837–1842 (1968).

157 QUENSEL, U.: Zytologische Untersuchungen von Ergüssen der Brust- und Bauchhöhlen mit besonderer Berücksichtigung der karzinomatösen Exsudate. Part 1. Acta med. scand. *68:* 458 (1928).

158 QUENSEL, U.: Zytologische Untersuchungen von Ergüssen der Brust- und Bauchhöhlen mit besonderer Berücksichtigung der karzinomatösen Exsudate. Part 2. Acta med. scand., suppl. 23 (1928).

159 RATZER, E.R.; POOL, J.L., and MELAMED, M.R.: Pleural mesotheliomas. Clinical experiences with 37 patients. Am. J. Roentg. *99:* 863–880 (1967).

160 REAGAN, J.W. and NG, A.B.P.: The cells of uterine adenocarcinoma (Williams & Wilkins, Baltimore 1965).

161 ROBBINS, S.L.: Pathology, pp. 941–943 (Saunders, Philadelphia 1967).

162 ROBBOY, S.J. *et al.:* Ovarian teratoma with glial implants on the peritoneum. An analysis of 12 cases. Human Pathol. *1:* 643–653 (1970).

163 ROBERTS, G.H.: Diffuse pleural mesothelioma. A clinical and pathological study. Br. J. Dis. Chest *64:* 201–211 (1970).

164 ROBERTS, G.H. and IRVINE, R.W.: Peritoneal mesothelioma. A report of 4 cases. Br. J. Surg. *57:* 645–650 (1970).

165 ROBERTS, G.H. and CAMPBELL, G.M.: Exfoliative cytology of diffuse mesothelioma. J. clin. Path. *25:* 577–582 (1972).

166 ROGOFF, E.E.; HILARIS, B.S., and HUVOS, A.G.: Long-term survival in patients with malignant peritoneal mesothelioma treated with irradiation. Cancer *32:* 656-664 (1973).

167 RONA, A.; MARSHALL, K., and RAYMONT, E.: The cytologic diagnosis of an ovarian mucinous cystoma from a virtually acellular specimen of abdominal fluid. Acta cytol. *13:* 672–674 (1969).

168 ROSE, R.L.; HIGGINS, L.S., and HELGASON, A.H.: Bacterial endocarditis, pericarditis, and cardiac tamponade. Am. J. Cardiol. *19:* 447–451 (1967).

169 SALYER, W.R.; EGGLESTON, J.C., and EROZAN, Y.S.: Efficacy of pleural needle biopsy and pleural fluid cytopathology in the diagnosis of malignant neoplasm involving the pleura. Chest *67:* 536–539 (1975).

170 SANDBERG, A.A.; YAMADA, K.; KIKUCHI, Y., and TAKAGI, N.: Chromosomes and causation of human cancer and leukemia. III. Karyotypes of cancerous effusions. Cancer *20:* 1099–1166 (1967).

171 SARRAZIN, A.; ROHELEC, G. LE et BOUSQUET, O.: Pleurésie et syndrome péritonéopleural révélant une affection pancréatique chronique. Bull. Soc. Med. Hôp. Paris *116:* 1209–1216 (1965).

172 SAUTER, C.: Diagnostische Bedeutung von Zellkulturen aus Pleuraergüssen und Aszites. Schweiz. med. Wschr. *101:* 1245–1247 (1971).

173 SCHILLER, W.R. and LAVOO, J.W.: Diagnostic peritoneal lavage in acute abdominal problems. Ohio St. Med. J. *70:* 313–316 (1974).

174 SCHINDLER, S.C.; SCHAEFER, J.W.; HULL, D., and GRIFFEN, W.O., jr.: Chronic pancreatic ascites. Gastroenterology *59:* 453–459 (1970).

175 SCHLIENGER, M.; ESCHWÈGE, F.; BLACHÉ, R. et DEPIERRE, R.: Mésothéliomas pleuraux malins. Etude de 39 cas dont 25 autopsies. Bull. Cancer *56:* 265–308 (1969).

176 SCHMITT, W.; PIETSCH, P. und TROEGER, H.: Die intraperitoneale Antibiotika-Spüldrainage bei diffuser Peritonitis. Zentbl. Chir. *97:* 3–10 (1972).

177 SERVELLE, M.; SOULIÉ, J.; TRICOT, J.F.; RAZAFINOMBANA, A.; CORNU, C. et ANDRIEUX, J.: Les épanchements des cardiaques. Archs Mal. Cœur *62:* 1119–1143 (1969).

178 SHAPIRO, B.S.: *Candida* peritonitis. Conn. Med. *30:* 727–728 (1966).

179 SHARMA, O.P. and GORDONSON, J.: Pleural effusion in sarcoidosis: a report of six cases. Thorax *30:* 95–101 (1975).

180 SHIFF, A.D.; BLATT, C.J., and COLP, C.: Recurrent pericardial effusion secondary to sarcoidosis of the pericardium. New Engl. J. Med. *281:* 141–143 (1969).

181 SIGUIER, F.; GODEAU, P. et HERREMAN, G.: Les péricardites sclérodermiques (à propos de trois observations). Bull. Soc. Med. Hôp. Paris *118:* 1299–1312 (1967).

182 SINGH, A. and KRISHAN, I.: Cardiac tamponade due to massive pericardial effusion in myxoedema. Br. J. clin. Pract. *24:* 347–350 (1970).

183 SMITH, J.P. and BORONOW, R.C.: Pseudo-Meig's syndrome with mucinous cystadenoma. Report of a case in which the effusions contained atypical cells. Obstet. Gynec., N.Y. *30:* 121–126 (1967).

184 SOCHOCKY, S.: Pericardial effusion. Laval Med. *39:* 539–546 (1968).

185 SOCHOCKY, S.: Chronic pericardial effusion. S. Dakota J. Med. *23:* 9–14 (1970).

186 SOLAN, K.: Amoebic pericarditis. Proc. Mine Med. Officers Ass. *45:* 20–22 (1965).

187 SOLOFF, L.A.: Pericardial cellular response during the postmyocardial infarction syndrome. Am. Heart J. *82:* 812–816 (1971).

188 SOOST, H.-J.: Lehrbuch der klinischen Zytodiagnostik, pp. 231–264 (Thieme, Stuttgart 1974).

189 SPRIGGS, A.I. and BODDINGTON, M.M.: The cytology of effusions, pp. 1–40 (Grune & Stratton, New York 1968).

190 SPRIGGS, A.I.: A simple density gradient method for removing red cells from haemorrhagic serous fluids. Acta cytol. *5:* 470–472 (1975).

191 STEPHANS, E. *et al.:* Heart and hypopituitary. 1. Constant electrocardiographic

anomalies 2. An unusual complication: pericardial effusion (apropos of 10 cases). Archs Mal. Cœur *58:* 1493–1502 (1965).

192 STOEBNER, P.; MIECH, G.; SENGEL, A. et WITZ, J.-P.: Notions d'ultrastructure pleurale. I. L'hyperplasie mésothéliale. Presse méd. *78:* 1179–1184 (1970).

193 SVANE, S.: Recurrent, hemorrhagic pleural effusion and eosinophilia accompanying pancreatitis. Acta chir. scand. *131:* 352–356 (1966).

194 SWENSON, N.L.; KUROHARA, S.S., and GEORGE, F.W., 3rd: Complete regression following abdominal irradiation alone, tne chylothorax complicating lymphosarcoma with ascites. Radiology *87:* 635–640 (1966).

195 TAKAHASHI, M.: Color atlas of cancer cytology, pp. 258–292 (Igaku Shoin, Tokyo/ Lippincott, Philadelphia 1971).

196 TAKEUCHI, J.; TAKADA, A.; HASUMURA, Y.; MATSUDA, Y., and IKEGAMI, F.: Budd-Chiari syndrome associated with obstruction of the inferior vena cava. A report of seven cases. Am. J. Med. *51:* 11–20 (1971).

197 TENG, C.Y.; NEMICKAS, R.; TOBIN, J.R., jr.; SZANTO, P.B., and GUNNAR, R.M.: Pericardial effusion following radiation to the chest. Report of a case. Dis. Chest *52:* 549–552 (1967).

198 TOMB, J.: A cytopathological study on serous fluid in cancer. J. Med. Liban *27:* 51–58 (1974).

199 TRAUB, Y.M. and ROSENFELD, J.B.: Malignant pheochromocytoma with pleural metastasis of unusually long duration. Chest *58:* 546–550 (1970).

200 VILLA, F.; GUZMAN, S. DE, and STEIGMANN, F.: Peritoneoscopic findings in ascites. Gastroent. Endosc. *15:* 48–50 (1968).

201 WEISS, K.: Zur Treffsicherheit der Tumorzellendiagnostik von Brust- und Bauch-höhlenergüssen. Med. Welt *24:* 1888–1892 (1973).

202 WHITE, H.A., jr.: Idiopathic total portal vein thrombosis with ascites. Alabama J. Med. Sci. *2:* 360–370 (1965).

203 WHITWELL, F. and RAWCLIFFE, R.M.: Diffuse malignant pleural mesothelioma and asbestos exposure. Thorax *26:* 6–22 (1971).

204 WIDAL, R. et RAVAUT, P.: Applications cliniques de l'étude histologique des épanchements sérofibrineux de la plèvre. C.r. Séanc. Soc. Biol. *52:* 648, 651, 653 (1900).

205 WIHMAN, G.: A contribution to the knowledge of the cellular content in exudates and transudates. Acta med. scand. *130:* suppl. 205 (1948).

206 WINK, K. und HAGER, W.: Der chronische idiopathische Perikarderguss. Dt. med. Wschr. *94:* 2160–2163 (1969).

207 WITTE, M.H.; WITTE, C.L.; DAVIS, W.M.; COLE, W.R., and DUMONT, A.E.: Peritoneal transudate. A diagnostic clue to portal system obstruction in patients with intraabdominal neoplasms or peritonitis. J. Am. med. Ass. *221:* 1380–1383 (1972).

208 WOLFE, W.G.; SPOCK, A., and BRADFORD, W.D.: Pleural fluid in infants and children. Am. Rev. resp. Dis. *98:* 1027–1032 (1968).

209 WOYKE, S.; DOMAGALA, W., and OLSZEWSKI, W.: Alveolar cell carcinoma of the lung: an ultrastructural study of the cancer cells detected in the pleural fluid. Acta cytol. *16:* 63–69 (1972).

210 WOYKE, S.; DOMAGALA, W., and OLSZEWSKI, W.: Ultrastructure of hepatoma cells detected in peritoneal fluid. Acta cytol. *18:* 130–136 (1974).

211 YAM, L.T.: Diagnostic significance of lymphocytes in pleural effusions. Ann. intern. Med. *66:* 972–982 (1967).

212 YEH, T.J.: Endometriosis within the thorax: metaplasia, implantation, or meta-
     stasis? J. thorac. cardiovasc. Surg. *53:* 201–205 (1967).
213 ZACH, J.: Die Cytologie der serösen Höhlen. Internist *11:* 401–412 (1970).
214 ZIPF, R.E., jr. and JOHNSTON, W.W.: The role of cytology in the evaluation of peri-
     cardial effusions. Chest *62:* 593–596 (1972).